WITHDRAWN

Oceans of Energy

Oceans

Reservoir

Augusta Goldin

of Energy

of Power for the Future

Illustrated with photographs and diagrams

Harcourt Brace Jovanovich / *New York and London*

Requests for permission to make copies of any part of
this work should be mailed to:
Permissions, Harcourt Brace Jovanovich, Inc.,
757 Third Avenue, New York, New York 10017

The excerpt from Homer's *The Odyssey* is reprinted
by permission of Penguin Books Ltd. from p. 174 of
The Odyssey, translated by E. V. Rieu
(Penguin Classics, 1946), Copyright © E. V. Rieu, 1946.

Printed in the United States of America

Library of Congress Cataloging in Publication Data

Goldin, Augusta R
Oceans of energy.

Bibliography: p.
Includes index.
SUMMARY: Discusses the possibilities of generating
electrical power from the world's oceans and of
extracting natural gas from ocean kelp farms.
    1. Ocean energy resources—Juvenile literature.
[1. Ocean energy resources.   2. Power resources]
I. Title.
TJ163.23.G64   333.9'14   79-3767
ISBN 0-15-257688-6

Videocomp-CRT Gael
First edition
B C D E

To Dr. Joseph Barnea,
friend, mentor, and critic,
with affection and
admiration

Contents

Acknowledgments

Many of the research workers currently investigating ways of extracting energy from the ocean—scientific visionaries all—have generously shared their expertise with me. All answered my numerous questions. Some checked and even rechecked the various chapters. Others accompanied their comments with the splendid photographs you'll find in this book. And that's how *Oceans of Energy* came to be written.

For their invaluable help, my gratitude goes to the following people whom I interviewed in New York and California: Raleigh Guynes, Defense and Space Systems Group, TRW Inc.; Dr. Thomas Henkel, Director of the Solar Energy Demonstration Project, Wagner College; Dr. Peter B. S. Lissaman, Program Director for the Coriolis Project and president, AeroVironment; Dr. Wheeler North, professor, California Institute of Technology, and marine biologist, Kerchoff Laboratory; Harold Ramsden, marine engineer, Global Marine Development Inc.; Walter Schmitt, oceanographer, Scripps Institution of Oceanography; Dr. Howard A. Wilcox, engineering physicist, Naval Ocean Systems Center; and Dr. Sidney Loeb, professor of chemical engineering, Ben-Gurion University of Negev, Israel, whom I interviewed in Montreal, Canada.

Gratitude, too, goes to the following for their assistance by way of telephone interviews: Carol Barth, Cosanti Foundation; Dr. William B. Cutler and C. Turner Joy, Jr., Lockheed Missiles and Space Company, Inc.; Victoria Edwards, Sovfoto/Eastfoto Agency; Octave Levenspiel, Oregon State University; Robert Radkey, AeroVironment; Paolo Soleri, arcologist, Cosanti Foundation; Dr. Harris B. Stewart, Jr., National Oceanic and Atmospheric Administration, Atlantic Oceanographic and Meteorological Laboratories.

With no less appreciation, I recognize the following for professionally helpful correspondence: Dr. Bruce H. Adee, Director, Ocean Energy Program, University of Washington; Dr. Richard Anderson, consulting geologist; A. Becker, Sea Technology; Dr. Charles E. Behlke, Dean, School of Engineering, University of Alaska; Sir Christopher Cockerell, Director, Wavepower Ltd., Southampton, England; Dr. Robert Cohen, Chief, Ocean Systems Branch, U.S. Department of Energy; Timothy D. Dellinger, Sea Solar Power, Inc.; Dr. Arthur Fisher, *Popular Science;* Julio Hernández Fragoso, Puerto Rico Water Resources Authority; Dr. Evan J. Francis, Director, Applied Physics Laboratory, Johns Hopkins University; Eileen R. Gull, Billings Energy Corporation; Barbara T. Hall and Mary C. Holliman, *Sea Grant '70s*, Virginia Polytechnic Institute and State University; Joseph L. Ignazio and W. F. Mackie, New England Division, U.S. Army Corps of Engineers; Nancy Jamison, Re-entry & Environmental Systems Division, General Electric Company; Archie M. Kahan, Office of Atmospheric Resources Management, U.S. Department of the Interior; D. C. Lefebvre, French Engineering Bureau; Dr. J. Leishman, National Engineering Laboratory, East Kilbride, Glasgow; Dr. William H. MacLeish, *Oceanus,* Woods Hole Oceanographic Institution; Dr. Michael E. McCormick, U.S. Department of Energy and director of ocean engineering, U.S. Naval Academy; Spark M. Matsunaga, U.S. Senator from Hawaii, Committee on Energy and Natural Resources; L. H. Miller, Hydraulic Research Station, Wallingford, England; Stanton S. Miller, *Environmental Sci-*

ence and Technology; Dr. Anthony Peranio, Technion, Israel Institute of Technology; Dr. Oswald A. Roels, Marine Science Institute, University of Texas; Dr. Stephen Salter, University of Edinburgh; Jack Schneider, U.S. Department of Energy; David Shapiro, OTEC Technical and Management Administration, Tracor Sciences and Systems; Dr. T. L. Shaw, University of Bristol, England; Janet Sillas, Brookhaven National Laboratory; John J. Smiles, National Oceanic and Atmospheric Administration, U.S. Department of Commerce; Roy R. Taylor, Engineering Technology Support Unit for the Department of Energy, Oxfordshire, England; Steven Tigemann and R. A. Whitehurst, U.S. Coast Guard; Dr. William Von Arx, Woods Hole Oceanographic Institution; Dr. W. W. Wayne, Stone and Webster Engineering Company.

For reading the manuscript in its entirety and providing helpful comments and suggestions, special thanks go to Dr. Joseph Barnea, Dr. Octave Levenspiel, Walter Schmitt, and Dr. Howard A. Wilcox.

Finally, my thanks to the librarians of the Engineers' Library, New York City; to the Hadlands, Harry, Harry Jr., and Karen; and a scroll of honor to my husband, Oscar, who, in more ways than one, is the driving force behind my personal energy.

If, despite so much assistance, errors still survive in the pages of this book, the fault rests entirely with me. And if I inadvertently omitted the names of any persons who helped me, my sincere apologies.

Augusta Goldin

You must take a well-cut oar and go on till you reach a people who know nothing of the sea and never use salt with their food, so that our crimson-painted ships and the long oars that serve these ships as wings are quite beyond their ken. And this will be your cue—a very clear one, which you cannot miss. When you fall in with some other traveller who speaks of the 'winnowing fan' you are carrying on your shoulder, the time will have come for you to plant your shapely oar in the earth. . . .

—The Odyssey, *translated by E. V. Rieu*

Oceans of Energy

1

The Ocean Is Charged

The ocean is charged with extraterrestrial energy that is not of the earth. It is energy that flows from outer space. It is safe, non-polluting, inexhaustible, and free.

From outer space comes the extraterrestrial energy of the sun. It warms the air and speeds the winds that drive the waves. It warms the ocean, which then stores thermal energy. And it propels the currents, which are simultaneously deflected by the spin energy of the earth.

Also from outer space comes the energy of solar and lunar gravity. This is the energy that creates the momentum of the earth-moon system and causes the tides.

The ocean is not a flat, lifeless body of water, but a super storehouse of restless energy. Here are the tossing waves, the surging tides, the crisscross currents, all brimming with energy.

Already wave-powered buoys and lighthouses dot Japanese waters. For years up-and-down wave motion has operated whistle buoys for the Coast Guard of the United States. Today, there's hard-

vith Extraterrestrial Energy

ly a coastal area in the world without a local inventor working on a wave machine.

Since 1966, two French towns have been completely electrified by tidal energy. At the Rance installation in Brittany, twenty-four reversible turbogenerators harness the tides as they come and go. The installation has an output of 240 megawatts. It is one of the most powerful hydroelectric stations in France.

This tidal power was free right from the start, but because the installation was costly, the price of Rance electricity was just a bit steep. In the 1970s, however, the picture changed. Every time the oil barons in the Middle East, Africa, and South America raised the price of oil, tidal power became more attractive because it became more cost competitive with the fossil fuels. Immediately, then, the Russians, the South Koreans, and the English turned to their coastlines and their drawing boards. They began thinking tidal and wave power, talking budget, and planning feasibility studies.

More recently, several ocean scientists were watching the Gulf

Stream sweep past the Florida coast at a lively five miles an hour. The idea of plugging into that warm water was intriguing. But could it be done? Could giant turbines and underwater propellers (resembling windmills) snatch energy out of the stream and generate electricity?

"It *can* be done," was the conclusion reached by the MacArthur Workshop held by the National Oceanic and Atmospheric Administration in Miami, Florida, in 1974. All agreed that there were problems, but none that money couldn't solve, because there is "nothing in such a project that's beyond the capabilities of modern engineers and technologists."

One particularly foward-looking scientist went so far as to predict that Gulf Stream electricity may well become cost competitive with other forms of energy in the 1980s.

There is also a magnificent life-support system in the ocean that is charged with gases and nutrients, salts and other minerals. In this system, dissolved oxygen sustains the sea animals from the smallest to the largest, from the amoeba to the shark. And dissolved carbon dioxide sustains the sea plants from the single-celled diatom to the towering 200- and 300-foot (60- and 90-meter) kelp.

For marine biologists, it was only a step from looking at the ocean as a natural life support system to trying to tap the energy of that system, deliberately and scientifically.

Backed by the United States Navy, in the mid-1970s a team of marine scientists, ocean engineers, and divers installed the world's first ocean energy farm 40 feet (12 meters) *below* the sunlit waters of the Pacific Ocean off San Clemente Island. It was a small farm. It was experimental in nature. It was planted with giant California kelp.

According to the project's director, Dr. Howard A. Wilcox of the Naval Ocean Systems Center at San Diego, California, "Up to 50 percent of the energy in this kelp can be turned into fuel—that is, into methane, which is natural gas. A future ocean kelp farm of, say,

Wave-powered bell buoy (Official U.S. Coast Guard Photo)

100,000 acres (40,000 hectares) might provide enough energy to supply an American city of 50,000 residents."

The ocean has always been threaded with the energy of waves and tides and currents. In ancient times, when fishermen watched the moving waters, they didn't know about "tidal energy" or "kelp farming," but they did know it was easy to ride their boats out on the ebbing tide and to bring back the catch on the incoming tide. They did know that the waves sometimes packed a fearful wallop, hurling stones up and over the seaside cliffs. They did know about "rivers in the sea" that would always carry them to certain islands. They also knew they could sustain themselves by living off the shellfish, the fish, and the edible greens that grew in the ocean.

Now, as the need for alternative fuels increases, oceanographers, chemists, physicists, engineers, and technicians are looking more searchingly at the ocean for hidden sources of energy.

There's all this salt dissolved in the sea. Could there be such a thing as salinity power?

There could be. The heavy concentration of salt is tempting a number of ocean scientists at the Scripps Institution of Oceanography in La Jolla and other centers to look into salinity power. To them it seems possible that salt-water/fresh-water batteries could be developed to produce large amounts of electricity.

Then there's the matter of temperature differences in the ocean. Between the Tropic of Cancer and the Tropic of Capricorn, the surface of the sea averages a warm and pleasant 82 degrees Fahrenheit (27 degrees Celsius). Two thousand feet (600 meters) below, the temperature drops to a dense and frigid 35, 36, 37, or 38 degrees F (2 to 3½ degrees C). Question: Is there a possibility that this temperature difference of some 45 degrees F could actually be harnessed to produce energy? Could a heat engine, floating beneath the waters, generate electricity?

Yes, this, too, could be done.

In the late 1920s, Georges Claude, a creative, determined, and very stubborn French physicist, decided to investigate this possibility. Choosing a site off the coast of Cuba, he managed, after a number of tries and almost as many disasters, to produce 22 kilowatts of electricity. This was a scientific breakthrough that brought hurrahs and brickbats from equally eminent scientists.

"With these warmed surface waters, with those lower colder water reservoirs, and with suitable technology, we're in business," crowed the proponents of ocean thermal energy. "According to our estimates, there's enough energy stored in the surface waters to supply 10,000 times the world's energy needs."

"Ha," countered the skeptics. "Georges Claude produced 22 kilowatts of electricity in Matanzas Bay. Was this net?" It was not. In order to get those 22 kilowatts out of his power plant, Claude had to put in 80 kilowatts to run his pumps. (Note: A kilowatt is equal to 1,000 watts.)

Today, Professor John Isaacs of the Scripps Institution of Oceanography calculates more conservatively. He estimates that with modern technology, power plants designed to generate electricity from the ocean's temperature differences could be made to produce twice as much electricity as the world uses at this time. This would be OTEC electricity, OTEC being the acronym for ocean thermal energy conversion.

This is a heartening prediction, but even if it came to pass, the results wouldn't solve all the world's energy problems. It would certainly be splendid to have access to a goodly supply of OTEC electric power, but (as of this time) electricity doesn't fly airplanes. It doesn't drive cars and trucks and buses. It doesn't propel ships across the sea.

Airplanes, cars, buses, and trucks can, however, be propelled by a gas that can be derived from water—and there's plenty of water in the sea. This gas is hydrogen, and it can be burned as a fuel. Hydrogen is the most abundant atom in the universe. In the ocean, it's locked up in every drop of water. Remember the formula for water? It's H_2O, meaning there are two atoms of hydrogen and one atom of oxygen in a molecule of water. Once hydrogen is extracted from water, it can be burned as a fuel and used not only to drive motorized vehicles but also to generate electricity.

A growing number of chemists and engineers are enthusiastic about a "hydrogen economy" in the future because, once hydrogen is extracted or in hand, it can be stored: as a compressed gas in tanks or as a liquid in cryogenic containers at minus 423 degrees F (minus 203 degrees C). It can also be stored as a solid after it is combined with an iron titanium alloy or magnesium to form metal hydrides. Then they can be transported and used as needed.

As long ago as 1847, the French writer Jules Verne, who was ahead of his time, envisioned just such a hydrogen economy. In his book *The Mysterious Island,* he predicted that people would one day employ water as a carrier for fuel. "Water," he wrote, "will furnish an inexhaustible source of heat and light."

Since Verne's time, ways have been found to extract hydrogen from water. One of the most successful is a process called *elec-*

Riding in a test hydrogen-powered vehicle
(Brookhaven National Laboratory)

trolysis. (An electric current is passed through water, and chemical decomposition takes place. Hydrogen gas and oxygen gas are liberated, and the liquid water disappears.)

In the 1960s, the National Aeronautics and Space Administration so successfully electrolyized water and so efficiently collected the liberated hydrogen that on the Apollo flights, hydrogen propelled the rockets and activated the fuel cells, and on the moon flights and in Skylab, hydrogen provided all-purpose electricity.

Thus, in the ocean, which covers 71 percent of the planet's surface, we have various kinds of potential energy that can be converted to basic fuel:

• There's the brute force of the waves and tides.

• There's the chemical force in the interacting gases, nutrients, salts, and other minerals in the water.

• There's the slumbering force of hydrogen locked away in the water molecules.

• There's the quiet mobile force of the currents, moving unceasingly in complex patterns through the ocean.

• There's the surprising force available in the temperature differences between the warmer surface and the colder depths.

So much energy in so many diversified forms means that in the future the world will not need to go energy-hungry. It will not have to be altogether dependent on any one or two basic types such as the old fossil fuels and the newer controversial nuclear fuels. Furthermore, the millions of villages near the coasts that don't now have access to power grids could, with diversification, provide improved living conditions for the inhabitants.

Those who live near rugged wave areas could develop and use wave energy.

Those who live near coastal inlets where the water, at high tide, advances like a roaring wall, could develop tidal power stations.

And for the rest of us, some of the ocean's energy could be converted offshore to methane gas, hydrogen, or electricity and then transported to the mainland by cable or ship.

All this energy has been swirling in the ocean since time began. And if we don't use it, it simply goes to waste.

Of course it isn't easy even to think of switching from the familiar traditional fuels—from oil and coal and natural gas—to the unfamiliar alternative fuels.

Thermal difference? Hydrogen gas, metal hydrides, and ocean energy farms? To many people these terms sound like science fiction.

And yet, though the extraction of ocean energy is still in the experimental stages, and production is still limited and costly, the fact remains that, with research and technological development, the energy of the future may to a considerable extent be drawn from the sea. When that will be depends on how soon the necessary processes become cheap enough. In the final analysis, it's not just the availability of free forms of energy in the ocean—it's the cost of extracting this energy that will determine how much of any one source will be developed.

Whenever that time comes, a changeover to ocean energy will carry a double bonus: first to the public's pocketbook, and second to the actual viability of Planet No. 3, our earth.

The pocketbook was first hit by the fossil fuel crunch in 1973. The price of oil soared because oil is basic to twentieth-century living, agriculturally, industrially, and residentially. An inflationary spiral was accordingly set in motion, and since research and experiment are costly, the search for new sources of fuel drove prices still higher.

Since fossil fuels are not inexhaustible, we are urged to conserve them and to increase our energy supplies by supporting the construction of nuclear power plants, which are expensive to build and hazardous to live with. Of course consumption would drop if we conserved. The United States, which has 5.3 percent of the world's population and uses 35 percent of the world's total of fossil fuels and hydroelectric power, might well reduce its consumption to 32, to 30, perhaps even to 25 percent. Some people contend that in all fairness, the United States should cut its consumption of energy down to 5.3 percent. But let's be realistic. The United States uses enormous amounts of energy to grow the grain that feeds a large portion of the world, as well as to manufacture the household appliances, cars, trucks, planes, and armaments that other countries clamor for.

Conservation, however, is only one part of the story. The other part concerns the Third World, which is racing to catch up with the standard of living of the industrialized countries, a standard that is based on energy. Today the Asians, Africans, and South Americans are moving away from a society based largely on muscle power. They are moving to develop industries, to build great jet fleets and massive war machines. Young people in those countries want jobs, modern jobs in motor mechanics, construction, aviation, electronics, and nuclear physics. This move is very real. It is fairly peaceable so far but likely to become warlike if a fair share of inexpensive energy is not made available to them.

If this push for equitable, inexpensive energy for all countries succeeds, perhaps as much as a thousand times the energy now used by the world will be required, and the *biosphere will not be able to handle the wastes from conventional fuels.* Nevertheless, Chauncey Starr, president of the Electric Power Research Institute in Palo Alto, California, holds that "one must assume that world energy consumption will move in that direction as rapidly as political, economic, and technical factors will allow."

As competition for the dwindling fossil fuels increases, pocketbooks will get flatter. They'll get flatter still as we battle the pollution of air, water, and heat created by the burning of the fossil fuels and the increasing number of cooling towers.

But why worry about developing new sources of fossil fuels? Why argue about the building of nuclear installations? The ocean is charged with energy that is clean, safe, and inexhaustible. It is there—waiting to be developed. That's bonus number one.

Bonus number two may turn out to be the continuance of our planet as a safe and livable place. The alternative, however, the increased use of fossil and nuclear fuels, might, according to some scientists, spell thermal trouble for the planet because increased use of these fuels could add two deadly and cumulative pollutants to the atmosphere—carbon dioxide and waste heat.

"Nonsense," sneers Joe Doaks. "We're constantly improving the air filters and scrubbers. Another year or so and factory smokestacks will be squeaky clean." And Jane Doaks adds smugly, "Aren't we cleaning up the car exhausts? Soon you won't even know the meaning of sulfur dioxide vapor. No need to worry about pie in the sky—I mean in the ocean."

There is, nevertheless, great need to worry about these two sneaky pollutants, carbon dioxide and waste heat, which are dispersed into the air from the smokestacks of fossil-fueled factories and other industrial plants, as well as some large apartment complexes.

But who would notice the increased carbon dioxide in the air? It's a colorless gas and odorless. It's the stuff that creates bubbles in soft drinks. And who would notice the long-range increase in global temperatures of one, two, or three degrees F?

The planet would notice because carbon dioxide would, in the course of time, form a blanket around it—a blanket that would prevent waste heat from radiating off into space.

According to Jacques Cousteau, ocean scientist and explorer, "When this carbon dioxide level passed a certain point, the 'greenhouse effect' would come into operation." This means that the heat radiating from the earth would be trapped beneath the stratosphere. This trapped heat would then raise the global temperature. Even an increase of one, two, or three degrees F would cause the glaciers to melt. Millions of tons of meltwater would then raise the level of the seas by as much as 200 feet (60 meters). The cities along the coast and in the major river valleys would be inundated— drowned and covered with water.

On this subject, as on many others, there are two camps of scientists. One camp sees the thickening blanket of carbon dioxide (CO_2) as the cause of heat entrapment that will lead to glacial melting— sees what Dr. Howard Wilcox calls the makings of a "Hothouse Earth." The other camp believes that the same thickening blanket of CO_2 will insulate the planet, prevent the sun's radiant heat from reaching us, and, consequently, bring on an ice age. Even our government is concerned. In an unreleased study, the Department of Energy warns that an increase in global temperature and CO_2 could change the world's climate and, eventually, create severe environmental and even political problems.

So, on a global front, what are we to do? Shall we dig deeper for the dwindling fossil fuels, build more nuclear plants—and risk changing the atmospheric temperature? Or shall we turn to the ocean, that storehouse of renewable energy, and consider ways of extracting this energy for our use?

In this book, only six of the ocean's energy sources are discussed: the tides, waves, and currents, the thermal and chemical differences, and kelp farming in the open sea (plus a word about hydrogen gas). At this time, these seem to be on the cutting edge of development. But there are additional sources which, in the future, may turn out to be very important. These include the exploitation of the sea bottom: for geothermal hot spots and for oil and natural gas, as well as for heavy crude oil, and gas hydrates. And not to be overlooked is the energy that could be obtained by harnessing the winds that sweep across the thousands of miles of open water.

Important for the future is man's determination to supply himself with the energy he needs. In various forms, much of this energy is available in the ocean and can be converted to his use with technological know-how. Already, there are thousands of studies in progress. Some will be doomed to failure, some fated for glittering success. As ocean scientists see it, only the glittering successes will provide abundant energy that is efficient, economical, and environmentally safe.

There is no time to lose. A sense of urgency focuses all eyes on the hundreds of millions of young people lined up to enter the labor market in every country and on every continent. The labor market operates on energy, but whether or not this energy is available, the young people will nevertheless keep coming. Their numbers will keep increasing, and their expectations will keep rising.

And all this while, the ocean is being charged and recharged with extraterrestrial energy, had we but the wit to use it.

2

What Energy Is

Since this book is about energy that is stored in the sea, let's talk first about energy in general: what it is, what it can do, and where it can be found. Let's also consider the conditions under which energy might be extracted and used.

In our industrialized society, energy makes the world go round. It spins the wheels on cars and boosts rockets into space. It toasts our bread and boils our eggs. It heats and air-conditions houses, lights up streets, and drives ships across the sea.

People may tell you that energy is oil and natural gas. Not so. These commodities are fuels. They must be burned before they can release energy, and the same is true of gasoline and coal and wood.

Scientists will tell you that energy is the capacity for doing work, and work is done only when a physical force (such as pressure or gravity) moves an object. To put it mathematically, work is force times the distance an object is moved. To put it simply, work is energy-in-action.

Have you ever seen a lid jiggling on a coffeepot, a runaway skateboard speeding down a hill, a raft riding an incoming comber, or a breaking wave tumbling children on the beach? These are all examples of work, of energy-in-action moving an object.

It was heat-produced steam pressure that jiggled the lid on the coffeepot. It was gravitational energy that propelled the skateboard. It was wave energy that moved the raft and tumbled the children on the beach.

In our working world, energy is basic because without it there could be no work. But if energy is available and used, any object or converter will do work: sometimes constructively, at other times destructively. Even a grand piano can do work.

Think of a shiny concert grand being hauled up an exterior wall of a remodeled tenement house. As long as the men are pulling on the ropes, they are applying a force that makes the piano move. At this time, it's the men who are doing the work, not the piano. The piano is only acquiring more and more potential energy the higher it rises above the ground. When the piano finally reaches the fifth floor, it will swing there, idly, as long as the men control the ropes and the pulleys. But suppose those ropes broke. Immediately the force of gravity would take over, and immediately the potential energy that's now stored in the piano would be released. The piano would plunge down. It would flatten everything in its path. It would hit the sidewalk and smash. Although such a smash is an accident, it is nevertheless, an example of a piano doing work. It is destructive work, to be sure, but it is work just the same.

The world is full of energy that can be used for jobs of work. This energy may be stored in people and pets, in pebbles and plants, in fossil fuels and trees and air, in rivers and lakes. But it's the oceans (those vast encircling waters that cover all but 29 percent of the earth's surface) that are the greatest reservoirs of stored energy.

This energy can be tapped—not from the ocean as a whole, but from each of the various *energy forms* contained in it. Ocean scientists are particularly interested in the following energy forms:
• the moving tides, waves, and currents
• the stored heat in the water
• the chemical composition of the ocean
• the stored energy in seaweeds growing in the sunlit waters

Now it happens that tides, waves, and currents are always in a state of change: now faster, now slower, now calmer or more violent. So, too, does the ocean temperature change, and the ocean's chemical composition. The energy systems therefore cannot attain a calm and constant equilibrium. They are, rather, in continuous *disequilibrium*. This disequilibrium turns out to be the key to energy extraction.

Octave Levenspiel of Oregon State University, a chemical engineer turned college professor, says, "When a system is *not* in equilibrium, that's when you can get work from it. This is the basic principle that governs all methods of extracting energy: from spouses, fuels, oceans, whatever."

In the ocean, disequilibrium manifests itself in three basic differences: in elevation, in temperature, and in chemical composition. These differences make energy extraction possible.

A difference in elevation is readily recognized as a source of energy. Who hasn't felt the thump of shower water in the bathroom or watched the high tides and higher waves?

In some places, the difference between high and low tide is negligible. In other places, such as the Rance Estuary and the Bay of Fundy, the difference is a thundering 44 and 60 feet (13 and 18 meters) respectively. Engineers know how to manage this difference in elevation between high and low tide. They build dams to hold the rising water in reservoirs. When the tide ebbs, they release the water through a turbine, and the energy of the falling water turns the turbine that spins a generator. In this way, they convert some of the energy in the tides to electricity.

Engineers also know how to extract energy from the waves, whose height is never in equilibrium. They're developing wave pumps and sea pumps and experimental floating stations for the production of electricity. But whatever the model, whatever the design, it works only because the surface of the ocean ripples and rolls and surges and is *not* in equilibrium. (See photos on pages 18, 21.)

A difference in temperature as a source of ocean energy is a more mystifying thought. If you took a thermometer when you went swimming in the sea, you would find the surface water uniformly warm or cool, depending on the day. If, however, you decided to tread water, your toes would soon signal a difference. At 5 feet (1½ meters) below, the water is cooler. Now recall if you will, the temperature in the tropics: 82 degrees F (28 degrees C) on the surface but 35-38 degrees F (2-4 degrees C) 2,000 feet (600 meters) below. This difference of some 45 degrees F is another example of a system that's not in equilibrium.

If there were no temperature differences in the ocean, if the thermometer read the same for the surface as for the depths, the ocean would be in thermal equilibrium. Then no energy could be extracted from it and no work could be done.

Engineers are ready to take advantage of this difference. They have the technology now to float unique and ingenious power

plants in the water, produce electricity, and transport it to the mainland by cable.

So certain are engineers that temperature differences can be made to produce electricity that some even theorize about conditions in hell. As the story goes, hell must be isothermal. This means that the temperature there must be in equilibrium, the same from top to bottom and side to side. Now assuming that some engineers might possibly go to hell and that hell by definition is an uncomfortable place, hot and humid, well—if it were not isothermal, not in equilibrium, there would have to be a difference in temperature. And if there were a difference in temperature, a clever engineer would quickly take advantage of that difference. He would build an air conditioner and make the place pleasant, cool, and comfortable. (But hell is not supposed to be a pleasant place!)

There's yet another source of energy going to waste at a point of chemical disequilibrium. This happens at the mouths of rivers, where fresh water mixes with the salt water of the sea. Here, too, energy could be extracted and used to grind grain, produce electricity—whatever. The technology for this kind of operation is about the same as that used for desalting water, only it must be worked in reverse. There's just one catch—for large-scale extraction of energy at the mouths of rivers, efficient and inexpensive technology has yet to be developed.

There's general agreement that the energy stored in the sea is available for extraction. There's just as much agreement that such extraction is very costly—many say, prohibitively costly.

Question: How can this cost be reduced?

Answer: Cost can be reduced by the implementation of the laws

U.S. Marine helicopter launching a wave pump from Kaneohe Bay, Hawaii. A section of the 290-foot (90-meter) hose is clearly visible. (U.S. Navy)

that govern the behavior of thermal energy. These are the laws of *thermodynamics.*

Thermodynamics is a word derived from the Greek. *Thermo* means heat and *dynamics* means force in motion. This sounds complicated, but it can be stated quite simply: thermodynamics is the study of heat energy as it relates to other forms of energy.

The first law of thermodynamics states that *energy can be neither created nor destroyed.* It can be only changed, transformed, or converted from one form to another. Consider this example. There's a certain amount of chemical energy stored in a truckload of coal. Burn the coal, and that chemical energy will be converted to heat energy. Now use that heat energy to boil a tankful of water. The heat energy will be converted to steam energy. If that steam energy is used to turn a turbine and spin a generator, the steam energy will be converted to mechanical energy, which will be converted to electrical energy, which may be used to spin music records (or operate vacuum cleaners, activate typewriters, or cook dinner on the stove).

According to the first law of thermodynamics, we can account for all the energy that was originally stored in the coal by measuring how much went into heat, into steam, into mechanical motion, into electricity, and into music waves from the record player. We can also account for the amount of energy that was wasted at each step. In this way, the first law allows you to balance your energy account book.

Since ancient times, alchemists and magicians have tried to create additional energy with "perpetual motion machines." They tried; they labored; they built amazing contraptions. But each time they failed because they couldn't break the strict accounting system of that first law of thermodynamics—that the amount of energy in the world is fixed. None can be created and none can be destroyed.

A wave pump in action in a very moderate sea continuously discharges water to a height of about 13 feet (4 meters). (U.S. Navy)

One would expect to derive a certain degree of comfort from the operation of this law because if new sources of energy are constantly being discovered (especially in the ocean), how could there possibly be an energy shortage? Surely the energy crisis is a myth.

The energy crisis is not a myth. As a matter of fact, energy crises, in varying degrees, may be with us for all time. The reason?

The second law of thermodynamics states that *whenever you try to extract all the useful work in an energy system, you fail.* Some of the energy is always lost to waste heat or friction. This is unavoidable. It's the tax that man has to pay to nature at each and every energy conversion.

Here's one case in point. An automobile engine (even the best) now uses only about 20 percent of the chemical energy in gasoline. The rest goes to cooling water, friction, and the hot exhaust. In a word, if you run your car with 20 percent efficiency, you pay an energy tax of 80 percent. Future engines may do better, but there will still be a tax paid for some waste heat.

And here's another case which shows how much energy is lost because of waste heat each time it is converted from one form to another. Let's go back to that truckload of coal and say that you start with 100 units of available energy. Burn the coal and generate steam, and you are left with 60 units. Put the steam through a turbine, and you generate only 40 units of electricity. Transmit it through power lines to your house, and you're down to 35. Light your room and capture a bit of the light in a silicon solar cell, and you're down to one unit of available energy. If you then run your high fi with this solar cell, you end up with one-tenth of a unit of Beatle pleasure. That is quite an energy tax.

In the real world, the second law of thermodynamics always claims an energy conversion tax, and it's never zero. Still, there's no need to despair. Even if we have to pay energy taxes to nature and even if the worldwide demand for energy doubles again and again, the sea will always be there as a great reservoir of renewable energy. As long as the sun shines and the wind blows and the tides roll and the waves rise and the currents thread the water, so long will the ocean continue to be charged with energy because the sea itself is served by an inexhaustible source—the extraterrestrial energy that flows from outer space. But the challenge to find ways of extracting this energy economically and safely will not be met for another decade or more.

3

Extracting Energy fron

As the moon in the sky grows and diminishes, so do the tides in the ocean grow and diminish. Some of the ancients, however, believed that the tides were caused by an angel moving his foot in and out of the water. Present-day understanding has abandoned that idea in favor of scientific knowledge. Today, high-level committees often meet to consider the movement of the tides and to devise ways of harnessing their energy. Their aim? To produce low-cost tidal electricity.

A thousand years ago, meetings devoted to the extraction of tidal power would have been tantamount to sorcery. In those days people still believed that the earth was flat, and some even believed the tides could be controlled by royal fiat. Case in point—King Canute, who commanded the tides to stop, stand still, and recede.

Now Canute, sovereign of Norway, Denmark, and England, was a highly intelligent ruler. He knew that the tides couldn't be halted by his decree, but his courtiers insisted that their sovereign could command anything. So, with a twinkle in his eye, Canute donned

Moon, your horns point to the east;
Wax, be increased.
Moon, your horns point to the west;
Wane, be at rest.

—ANONYMOUS

he Tides

his kingly robes—crown, scepter and all—seated himself on a long-legged portable throne, and ordered those selfsame courtiers to carry him to the seashore when the tide was low. There they waited as the tide turned. There they waited as the waters rolled landward, the breakers rose, and the surf splashed the beach. Finally, when all the sand and all the courtiers were well drenched, Canute raised his scepter. "Stop!" he thundered from his high, dry, and cushioned seat. "Stop, you incoming waters, and recede."

Of course the incoming waters neither stopped nor receded. The tides that rock the oceans are beyond the command of mortals because they are controlled by a complex interplay of forces. There are, first of all, the two forces from outer space: the gravitational pull of the sun and of the moon. The sun, being 93 million miles distant, exerts the lesser pull, something like 46 percent of the moon's pull. The moon, being closer (close enough for astronauts to visit), exerts a more powerful one that is strong enough to create tides twice a day in most places of the world.

One factor that produces very high *spring* tides and very low *neap* tides each month is the position of the sun and the moon in relation to the earth. The rotational spin of the earth is another factor. Still another is geography—meaning that the shape of the oceanic basins and the coastal formations often create local conditions not to be duplicated anywhere else. This, for example, accounts for the 60-foot (18-meter) tides in the Bay of Fundy and the 1-inch (2½-centimeter) tides at Tahiti.

But in King Canute's time, this kind of scientific understanding was still far in the future. People who lived near the sea, however, had learned about the relationship between the moon and the tides. They had noticed that the moon rose fifty minutes later each day, and as the moon appeared later, so did the surging high tides. They had also noticed that the moon's appearance changed in the course of the lunar month (which is twenty-eight days). And all knew there were very high tides when the moon was a thread of silver in the sky and again when it was a full round glowing disk. They all knew as well that when the moon was at its first and third quarters, the range (or difference) between high and low tide was moderate. By following the moon-tide timetable, seafarers made a point of sailing with the tide because, as everyone knew, a low tide could easily ground a vessel, just as a high tide could float it free.

Millers along the coast had another good reason for watching the moon and the tides. It was the millers who first learned to harness the tides. They built tide mills and ground wheat and barley. In a word, it was the millers who found out how to make tides do the work of many men. They learned to convert tide power to mechanical energy, just as later engineers were to learn to convert mechanical energy to electrical energy.

By the eleventh century, tide mills were creaking and groaning and clattering along the shores of England. Wherever the tide reached sufficient range, there were the brawny millers, their hair, arms, and smocks all dusted with flour, clumping about in wooden shoes, tending the sloshing waterwheels, and seeing to the grating millstones.

An enterprising miller would build his mill at an estuary—the spot where a river meets the ocean. Here, together with his stalwart sons, he dammed the river with rocks and mud and logs. In this way, he created a storage pond behind the dam.

Special devices set into the dam controlled the water in the storage pond. These were simple automatic lock-type gates or flaps. When the tide rose, the flaps opened inward and the water flooded the pond. When the tide ebbed, the flaps closed. When it was needed, the water was released (through a narrow sluice gate) to turn the mill wheels.

Tide mills were good businesses, and there's a record of one at the port of Dover which operated so briskly that it interfered with harbor traffic. Today, most of the old mills are gone. Only one is still standing and still grinding grain in Suffolk, England. Naturally, it's a great tourist attraction since it operates exactly as it did in the twelfth century: with storage pond, dam, flaps, sluice gate, and four sets of millstones.

But Britain held no patents on tide mills. By the twelfth century, similar ones by the score thundered along the western coast of Europe, enriching the Bretons in France, the Dutch in Holland, and the Spaniards along the Bay of Biscay.

After the New World was discovered, the ships that carried the immigrants also carried the plans for tide mills. The idea of using tidal energy fascinated the early settlers. On the southeast coast of Canada, the French enlisted the help of the Indians, and built the first tide mill in the New World. The year was 1600.

Soon after, the colonists in New England also turned to tidal energy and built grist mills and mills for grinding spices and sawing lumber. By 1734, they had constructed a Rhode Island model, a four-wheel installation with 20-ton wheels that could generate an astonishing 50 horsepower.

Energy from the tides can be extracted in a number of ways. When our ancestors operated those ancient tide mills, they extracted *potential* energy that was stored in the water they had impounded in the storage ponds. Later, when efficient pumps were invented, they learned how to extract another form of energy from

Ancient Saint Suliac tidal
mill on the Rance River in
France (Phototheque EDF,
Habans, courtesy Electricite
de France)

the surging tides: kinetic energy, which is the energy of moving water.

The first large-scale extraction of energy from the moving tides came about in the second half of the sixteenth century. At that time Londoners were thirsty. Water carriers did their best, but local wells and springs, and even the fountains in the square, often ran dry. More often they ran polluted. Needed was a city water-supply system that was dependable and safe. Accordingly, a company was organized to pump water from the Thames River, which is washed by the tides. The water ran fast in this river where it narrowed between the massive piers that supported London Bridge. Here, on wooden frames that rose and fell with the tides, the company installed three wooden waterwheels. Now these were no ordinary waterwheels; they were designed to operate in either direction, depending on tidal flow. The company also installed new and efficient pumps—sixteen to each wheel. Then the tides turned the reversible waterwheels, the wheels drove the pumps, the pumps forced the water into the pipes, and the pipes supplied London with water.

To Londoners, the new waterworks proved that tidal energy could be utilized economically, for the public good. To scientists and engineers, it proved that tidal energy (both potential and kinetic) could be transformed to mechanical energy and used to do work.

Still, there were serious limitations to the use of tidal energy— limitations that seemed insurmountable. For one thing, it had to be extracted "in situ" and used immediately. For another thing, although the tides could turn millwheels well enough, the millers had to work strange hours. Since the moon and the high tides rose 50 minutes later each day, an eager beaver miller would, periodically, have to operate his mill at night. Still, this was free energy, so despite these limitations, tidal energy continued to intrigue coast dwellers.

This fascination peaked in the twentieth century. By that time, the engineers had learned not only how to transform tidal energy to mechanical energy, but they had also learned how to transform mechanical energy to electrical energy.

The time was right for the construction of a tidal hydroelectric power station. Such a station would be backed by a huge storage basin or reservoir. It would operate with reversible turbines and pumps. It would run on complex operational cycles. And such a station could extract energy from the tides both at flood time and at ebb. Such a station would transform tidal energy to mechanical energy to electrical energy. It would tie in with a grid and supply electric current to people and places hundreds of miles away. And it would be pollution free.

Today, the world's first and largest tidal power plant stands on the coast of Brittany, overlooking the turbulent English Channel. Of twenty-two possible sites, this one, at the mouth of the Rance River, was selected by the French engineers for very good reasons. The lo-

A map showing the location of La Rance Tidal Power Station (Electricite de France)

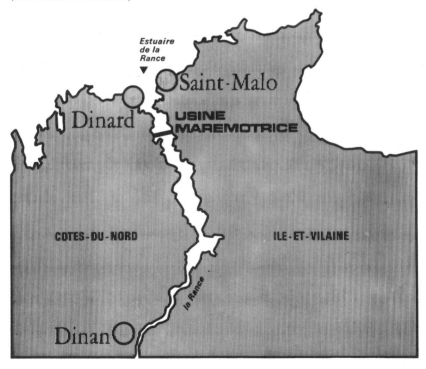

cation for the reservoir was ideal. The tide performed admirably, with a flow in and out of the estuary of 630,000 cubic feet (18,900 cubic meters) of water per second. The range topped 44 feet (13 meters) and the speed of the tide (between Brest and St. Malo) was often clocked at 55 miles or 90 kilometers an hour. Not surprisingly, motorists sometimes try to race the incoming tide, but they often come off second best.

Of course La Rance wasn't built in a day. It was talked about, re-searched, and studied for half a century. First the turboalternators had to be invented. Then the technology had to be developed for building an installation of such magnitude. And not to be over-looked was the fact that the budget called for something close to $100,000,000—a tremendous sum for that time. Finally, in 1959 the project design was completed.

In 1961 the project was started—2 miles (slightly over 3 kilome-ters) upstream from the mouth of the estuary where the channel is 2,500 feet (770 meters) wide.

No sooner was ground broken than tremendous enthusiasm swept the country, and the media took over. The French newspapers car-ried screamers calling the project "Une Grande Réalisation Inédite" . . . an unprecedented achievement.

The French radios cheered. "A revolutionary project. The world's first. We are producing French-style electricity . . . tidal electricity that is virtually independent of the moon's timetable . . . that can operate in a rhythm more in keeping with human activities."

The French television stations beamed pictures of men and ma-chines transforming a pastoral river scene into an industrial master-piece. For six years they televised the hardhats riding clattering bulldozers, roaring backhoes, creaking cranes, and rotating cement mixers. They showed geologists verifying the borings of rock forma-tions, hydraulic engineers studying the intensity and direction of tidal flow, and architects (blueprints in hand) checking the installa-tions of caissons. They zoomed their cameras in on the piles being driven into the stream bed and panned to the cut-off cofferdam and the reservoir.

Finally in 1967 the Rance Tidal Power Station was completed.

La Rance Tidal Power Station under construction (Phototheque EDF, Brigaud, courtesy Electricite de France)

Dam, power plant, navigation locks, and sluice gates were all in place. The large single storage reservoir brimmed with water.

Topside, the dam now carried a double roadway, crowded with holiday cars rolling between two fishing and resort towns: walled St. Malo and Dinard. Within, the dam now housed twenty-four turbo-alternators (also called reversible turbines). Each one can operate as a turbine or a pump. Each one can operate in either direction, basinward or seaward. Each one faces downstream for the incoming

Aerial view of La Rance Tidal Power Station connecting the two towns of St. Malo and Dinard. See the electric power station in lower left. (Michel Brigaud, French Embassy)

tide; then, as the tide turns, the turbine blades are turned until they face the ebbing tide. Computer-programmed, La Rance has solved the problem of tidal irregularity, but not 100 percent, because output varies. Under optimum conditions, a generous 240 megawatts can be chalked up, but on a year-round basis, the average computes to 25 percent of rated capacity. This 25 percent may not sound superefficient, but it is comparable to intermediate load plants that burn coal or operate on number two fossil fuel oil.

In the future improved technology will no doubt boost that 25 percent, but as one of the engineers at the French Engineering Bureau noted, "Although the principles of tidal power generation are relatively straightforward, the efficient extraction of a *constant* amount of power is more complicated."

Almost before the world's congratulatory messages could be delivered to the French, the world's second tidal power station came on line. This was a small Russian pilot project at Kislaya Guba, near Murmansk, on the Barents Sea. Completed in 1968, it began producing a modest 400 kilowatts of electricity immediately.

The Russians developed Kislaya Guba for a special reason—to find out if certain innovative construction methods would work under the harsh conditions of the far north. If these methods worked, the Russians, who are practical dreamers, planned to develop scores of small tidal power stations along the White Sea for low-cost electricity. They envisioned new towns going up, new industries flourishing where none had flourished before, and a raised standard of living for millions of people yet to be born.

At Kislaya Guba, the tides are ample and predictable, but there were problems. That northern area was barren and completely uninhabited. To undertake the construction of a power station there, a viable community would first have to be established. Workers would have to be shipped in, houses raised, roads built, a marketplace laid out, and shops set up to handle the heavy equipment. To develop such a community would cost a fantastic amount of money. "It would be cheaper," suggested one committee of engineers, "to build the tidal power station elsewhere—in a port that already has

Tide-operated power station being prepared for its sea voyage from Prityka

Cape to Kislaya Guba (Tass from Sovfoto)

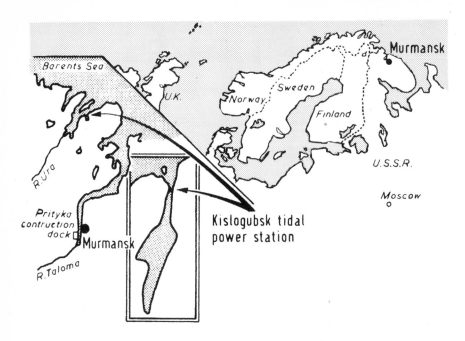

Site of the towed-in Kislaya Guba (spelled Kislogubsk in this map) tidal power station (*Chartered Mechanical Engineer*, 7/78)

workers, shops, tools, and material. Then all we would have to do is *tow* the completed station to Kislaya Guba."

No sooner said than done. In Prityka Cape, a more favorable port with good factory conditions, the Russian labor force got to work. Using a single reversible bulb turbine (such as the French used at La Rance), they created a prefabricated power station on a floatable caisson. This they towed to a barrage site at Kislaya Guba, sank it in the mouth of the gulf, tied it into position, and flanked it with prefabricated dam sections. Because they didn't have to build a cofferdam (a pumped-out enclosure under the water) and didn't have to establish a working village, the cost was nominal. But the achievement was tremendous. And exuberantly the Russians estimated that, in time, they would be able to open up the frozen Arctic for industrial activity and settlement.

In actuality, this project proved that it would be technically possible to build and tow even larger tidal power plants to distant and isolated sites. Based on his experience, Lev B. Bernstein, chief engineer at Kislaya Guba, had this to say in the April 1974 issue of *Civil Engineering—ASCE,* "The USSR is [now] designing for Mezen Bay on Russia's Arctic Coast, a tidal power station that will have a capacity of 6 million kilowatts. Another possible site is Tugu Bay at the

The tidal power station being towed to Kislaya Guba (Tass from Sovfoto)

coast of the Sea of Ohotsk. And a 20 million kilowatt station might be considered in Penginsk Bay."

Tidal power is eyed favorably in the U.S.S.R. The Russians know that even small tidal stations in the far north and northeast would provide electricity that would, otherwise, be unobtainable since the nearest electric grid is many hundreds of miles away.

While the French cheered La Rance and the Russians worked out their construction methods, the United States and Canada talked. At issue was the development of a huge hydroelectric station at Passamaquoddy Bay. The talk started in 1919, when the American engineer Dexter Cooper stood on the shores of Passamaquoddy Bay, which separates Canada from the state of Maine. Cooper was awed and impressed by the monstrous racing tides that move two billion tons of water a day. "Here," he declared, "I will build the world's

Proposed schematic showing the upper and lower basins at Passamaquoddy Bay (U.S. Army Corps of Engineers)

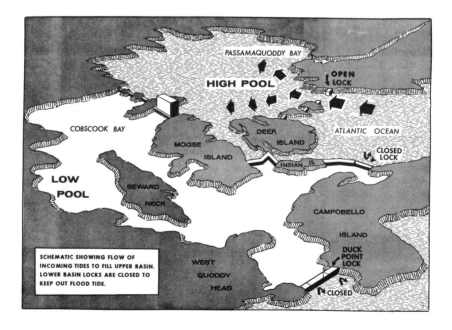

first tidal power plant." He envisioned two pools (or basins), using Passamaquoddy as the high pool and Cobscook Bay in Maine as the low pool, connected by a tidal power station that would generate electricity.

But nothing happened until 1935 when Franklin Delano Roosevelt, then President of the United States, entered the picture. FDR, who often spent his vacations sailing on "Quoddy," shared Cooper's vision. He allocated $7 million to the U.S. Army Corps of Engineers, ordered them to start construction, and shipped 5,000 men north to Quoddy Village, which was a hastily constructed mix of barracklike buildings. Their instructions? To develop a single storage basin at Cobscook Bay, to be linked with the Canadian basin at Passamaquoddy when that was built.

In one year, the 5,000 men completed three small dams—mere sections of what was to have been a large enclosure. Then they were recalled. The $7 million had run out, and Congress refused to allocate further funds. While the vote was being taken, there were near riots on the floor and boos and catcalls and shouts of "boondoggling" and "sell-out." But Congress stood firm. And the scuttlebutt was that the electric power companies frowned upon the development of tidal power.

Both Cooper and Roosevelt were ahead of their time. Talk was to continue for another forty years, with important commissions alternately voting yea and nay. Seemingly, the Passamaquoddy project was more of a political hassle than a scientific one.

By the mid-1970s, the need for abundant low-cost electricity had become a pinching necessity, so consideration of the project was reactivated. Once again the United States Army Corps of Engineers turned its attention to Cobscook Bay with its 18-foot (5½-meter) tides.

Four years later, the corps was ready to report the results of its $18 million study to the citizens of Maine. In July of 1978, public meetings were called. In Eastport and Augusta, the town halls were packed. Wiry farmers with weatherbeaten faces and gnarled hands, brisk housewives, and prosperous storekeepers wanted to know what the government was going to do for the Down-Easters. Many mumbled, "We're already paying 28 percent more for our electricity than the rest of the country."

Blue-jeaned co-eds, drop-outs, and veterans (many of them unemployed) pushed for a new day—one that would utilize an alternate form of energy. "We've got that energy right here. Cobscook Bay has the highest head (tidal range) in the continental U.S.A."

The corps reported the results of the study quietly, courteously, and sympathetically, saying that "(a) the project does not now appear to be economically viable in terms of benefits to costs and . . . (b) the project could be expected to pay for itself *provided* fuel costs escalate more rapidly than general inflation."

Across the bay, however, Canadian studies expressed greater optimism and confidence. A 1977 report, released by the Bay of Fundy Tidal Review Board recommended that "immediate consideration be given to the resolution of the financial constraints, to developing tidal power . . . because certain sites in the bay are now close to economic acceptance."

Three thousand miles to the west, Alaskans are also considering their tides and tidal power but without the urgency of the Down-Easters. Alaskans have been intrigued by their tides for more than two centuries, ever since Captain James Cook explored the inlet that now bears his name. Cook was so amazed by the extraordinary tides that he wrote about them in his *Journal:*

> Saturday, May 30, 1778 . . . Here we lay during the Ebb which ran 5 knots an hour.

> Sunday, May 31, 1778   At 9 o'clock we came to Anchor in 16 fathom . . . and found Ebb already made; which when at its greatest strength . . . fell upon the perpendicular, after we had anchored, 21 ft. . . .

> Monday, June 1, 1778 . . . the flood set strong into the River Turnagain, and Ebb came out with still greater force and the water fell upon a perpendicular, while we lay at anchor 20 ft. . . .

A tide gauge at Anchorage, Alaska. The photograph shows a range of 34 feet. (National Oceanic and Atmospheric Administration)

Ever since Cook's time, navigators and explorers and fishermen in Cook Inlet have observed those high, fast-flowing tides with awe. Now, in the last quarter of the twentieth century, this awe is turning into a dream for tidal development.

"A tidal power station," one is likely to hear in any bar and grill on a Saturday night, "even a little one, would spread cheap electricity around. Think what that would do for the state of Alaska!"

Alaskan engineers react to this kind of talk with professional reserve. They know Alaska is not plagued by the kind of political hassles that becloud Passamaquoddy and Cobscook Bay. But they also know that Cook Inlet has two unique problems that will have to be faced when the time comes: the problem of ice and the problem of the tide's peculiar timetable.

For five or six months of the year, Cook Inlet is choked with ice. Great sheets of ice, 3 to 6 feet (1 to 2 meters) thick, drift with the tides. Ice chunks, some as big as houses, form on the tidal flats where they melt and refreeze, melt and refreeze, and become studded with bottom sand and silt. And all the while, they roll and tumble with the tides and, together with the ice sheets, crash against the piers like battering rams.

Although there are a number of sites along the channel that could be developed as tidal power stations, no one minimizes the icing problem. According to Dr. Charles Behlke, dean of the school of engineering at the University of Alaska, "It would be difficult to prevent such ice chunks from damaging dams and power stations unless the whole project was positioned below the surface of the water and the floating ice."

Of course, the engineers could construct underwater power installations. They would tow in a prefabricated power plant to Knik Arm (the most favored site), add refilling gates in floatable prefabricated cells, plus submerged sluice gates, and without too much trouble the ice problems would be considerably reduced.

Small-scale power stations could also be developed at certain sites along the shores of the channel but again clever engineers would be needed. Here they would have to work with a peculiar tide. Behlke, who has studied the tides exhaustively, notes that there is a built-in time lag in the tide's timetable. This lag works out in such a way that when high tide races into the *mouth* of the channel, the people at the *head* of the channel, just 260 miles (416 kilometers) away, are experiencing an 11-hour lag—in other words, a low tide.

A hundred years ago, such tidal performance was considered mysterious, weird, and unpredictable. It's still unusual, but today computers can print out tide schedules at each of the favorable sites along the shores of the inlet. And should power plants be built, the computers could program the amount of water that would be available at each of the spatially lagged tidal storage basins.

With resourceful engineers, with money for more research, and

Low-tide ice build-up on the piling supporting the dock at Anchorage, Alaska (Dr. Charles Behlke)

with time, mighty tidal power stations can be expected to rise at selected channel sites, there to convert the kinetic energy of the extraordinary tides to electricity. "But such tidal power development is still in the future," says Dr. Behlke. "Even a small prototype station in Cook Inlet would cost $100 million to $200 million, and the local need is not yet so acute. We are a small state. Twenty to fifty years from now, such projects will no doubt become very attractive."

In the meantime, the thundering tides continue to sweep the inlet as an inexhaustible source of energy.

Conditions at the Rance River, Kislaya Guba, Passamaquoddy and Cobscook Bay, and Cook Inlet are unique. Few places in the world are so well provided with extraordinary high tides and fine estuaries or narrow channels that could be dammed for water storage. There are, however, scores of inlets on various continents with heads exceeding 15 feet (4½ meters) that could be developed as small-scale power stations. Such sites dot the coasts of British Columbia, Brazil, Argentina, Chile, the Straits of Magellan, and the Ivory Coast near Abidjan in Africa.

There are also spectacular tides that crash against Asian shores, where the need for low-cost energy is greatest because the per capita consumption of energy is the lowest in the world. Tides with heads of 40 feet (12 meters) or more surge into the Russian Sea at Okhotsk, into the mouth of the Seoul River, and along Shanghai, Amoy, and Rangoon. Great tides also favor India and run high in the Gulf of Ambay and the Gulf of Kutch.

After the Vietnam War ended, more than casual interest in tidal development began to simmer in South Korea and India, in Australia, Argentina, and England.

In barren northwest Australia, the Walcott Inlet, which empties into Collier's Bay, is being eyed keenly. The reason? Much energy will be needed to exploit newly discovered deposits of bauxite and iron ore.

In England, the sixty-year-long discussion concerning the development of the Severn Barrage (a tidal power station across the Severn estuary) is only now gaining urgency. Of the three major proposals, one calls for a single basin scheme, one calls for a double basin scheme, and one calls for a modified scheme. Most engineers favor the double basin scheme because it would be able to deliver electricity as needed, either continuously or intermittently, during peak demand.

As is true of most innovations, tidal-power development has its detractors. The initial cost of power installations is said to be high and the cost of electricity produced by these installations is also said to be high.

As a matter of fact, where conditions are favorable, these criticisms concerning cost are no longer valid. Prefabricated, floated-in caisson modules (for powerhouse and sluiceway sections) can now lower capital costs, as the Russians demonstrated at Kislaya Guba.

Also, inflationary concerns for the next fifty years need not be a cost concern because the useful life of such a project is estimated at fifty years or more.

And, most important, tidal electricity has become low-cost electricity. The Colson Research Symposium (meeting at Bristol University in England, in 1978) found that "Rance electricity is now cheaper than electricity produced by nuclear plants."

Environmental effects? In the vicinity of the dam, there is some damage to marine life as a result of a slight increase in water temperature and a reduction in oxygen content. Also, the dam interferes with fish migration. These, however, are relatively minor hazards, more than balanced by the attractive reservoir that serves as a fishing and recreational resort.

Understandably, some experts are betting on the everlasting tides as a future source of nonpolluting, low-cost electricity. Some are even predicting that long before the turn of the century, great tidal power stations will be producing that electricity abundantly, regularly, and consistently in areas where conditions are right. Other experts are looking at the ubiquitous waves that ride the global tides. They are predicting great wave-power projects. We'll take a look at some of these projects in the next chapter.

4

Extracting Energy from

Let someone say the word "ocean," and immediately people visualize waves: sun-streaked, foam-flecked, white-capped, or mountainous. There's energy in those waves—enough energy, say the ocean scientists, to electrify the world—and this energy is whipped into the water by the winds.

When the day is sunny and the wind gentle, the ocean is calm. The waves are low, then, with just enough energy to float things; rafts, sailboats, castaway bottles, bits of driftwood, or children strapped in life preservers. But let the sky turn leaden, the wind rise, and the waves will run before the wind. As the wind shifts and veers, so do the waves—racing from many directions, tumbling, rolling, breaking, their crests foaming and blowing away. Should the wind increase in violence, so will the waves. They'll grow taller and run faster. They'll absorb more of the wind's energy. And every time they double in height, they increase their energy fourfold.

When waves are high and violent, they're packed with lots of energy. Imagine waves strong enough to hurl a 139-pound (69½-kilo-

The rising world of waters, dark and deep
—MILTON, *Paradise Lost*

he Waves

gram) rock through the roof of a lighthouse. This happened at Tillamook Rock, off the coast of Oregon, and the lighthouse stood on a cliff 135 feet (40 meters) above sea level. In France, waves have tossed 3-ton (2,700 kilogram) boulders over Cherbourg's 20-foot (6-meter) seawall. In Hawaii, during storms, waves 100 to 200 feet (30 to 60 meters) high crash against the lava coast of northern Oahu—a mere commonplace in that area.

The explosive energy of the waves has been reported ever since mariners first set sail. In the nineteenth century, Sir John Murray, one of the world's greatest oceanographers, reported tremendous buffetings from mountainous seas. "I have," he wrote in his logbook, "observed waves towering up to 60 feet." But the highest wave ever encountered in mid-ocean, estimated at 112 feet (35 meters), was reported in 1933. On a moonlit night in February, the naval tanker U.S.S. *Ramapo* was running downwind in a rough Pacific sea. The winds, with a fetch (an unbroken sweep) of thousands of miles of open water, had built up to super-hurricane force. On the *Ramapo*'s bridge, the watch officer, soaked in his oilskins and clinging to the rails, was startled to see an incredibly large wave rising astern. As the ship's stern rode in the trough of the sea, he sighted through the crow's nest to the top of that monstrous wave. He calculated the angle at which the ship was riding. He checked the height of the crow's nest. The rest was simple geometry and computed to 112 feet, a height comparable to that of a twelve-story building.

In an indirect way, the tremendous energy contained in the waves is extraterrestrial. It comes from the sun, which drives the winds. The winds, in turn, transfer their energy across the air/water interface to drive the waves.

In parts of the world where the winds are weak, the waves are in

Violent waves hurled a 139-pound (69½-kilogram) rock through the roof of Tillamook Lighthouse, which stood on a cliff 135 feet (40 meters) above sea level. (Oregon Historical Society)

most instances low and gentle. But in the trade wind belts (both north and south of the equator) and in the North Pacific and around the stormy British Isles, wave action is vigorous and wave energy is significant. According to recent estimates, this energy is equal to about 30 percent of all the energy utilized in the world in the latter part of the twentieth century. Such estimates, understandably, spur the nations to invest in wave-energy research. The United States Department of Energy (DOE) recently suggested that, in the North Pacific alone, converted wave energy could produce 5 to 50 megawatts of electrical power per kilometer (⅝ mile) of coastline. Stephen Salter, physics professor at the University of Edinburgh, Scotland, calculates that a 300-mile (480-kilometer) chain of his energy extraction devices, moored in the Atlantic around the Hebrides, could convert enough wave energy to supply all the electricity now required in Great Britain.

Over the years, as wave data were collected and organized, ocean scientists began thinking about extracting wave energy for our use. But always an important question remained unanswered: How can we extract substantial amounts of this wave energy efficiently and economically?

So many people tried to answer this question that in England alone more than 340 patents for wave-powered generators were registered between 1856 and 1973. Most of these patents, especially the more recent ones, called for some type of float that rides the waves. In such a float, the up-and-down motion of the waves drives a rotating device, and this device converts the kinetic energy of the waves to mechanical energy. Simple? Yes, but all those patents remained on paper.

Not until the last quarter of the twentieth century were any effective projects set up—not until Britain's Department of Energy funded a number of active research studies. These, naturally, were located where the strongest waves roll: on the shores pounded by the Atlantic Ocean, the turbulent English Channel, and the North Sea. By 1979, there were four front-runners: the Salter nodding duck, the Cockerell raft, the oscillating water column, and the Russell rectifier.

The Great Wave at Kanagawa. Japanese print by Hokusai (Metropolitan Museum of Art, The Howard Mansfield Collection, Rogers Fund, 1936)

The Salter duck is the brainchild of Dr. Stephen Salter who, in addition to being a physics professor, is a mechanical engineer, a lecturer in artificial intelligence, and an inventor. Salter stumbled on his wave energy project quite by accident. As he explains it, "I caught the flu in 1973, and my wife said to me (with callous indifference to my misery), 'Stop lying there feeling sorry for yourself. Why don't you solve the energy crisis?' And what she wanted was something that would provide the vast amounts needed, would be clean and safe, would work in winter in Scotland, and would last forever."

Salter, the physics professor, started thinking about the tumultuous waves off the coast of Scotland. Salter, the engineer, did some sums and was amazed at the amount of wave power that seemed to be available. Salter, the inventor, saw that the obvious extraction device would be something like a lavatory ball cock, bobbing up and down, working a pump, and producing electricity. What he needed, then, was a dynamometer that would measure the work done by the bobbing ball cock, and a housing to hold the mechanism together. So, with balsa wood and glue, transistors, and a homemade dynamometer, he started inventing. When done, his creation looked like a teardrop with a beak. When floated in a borrowed wave tank, it bobbed up and down like a nodding duck—hence its name.

The early model was able to extract 15 percent of the available wave power. That was good and very heartening. An improved duck subsequently achieved up to 90 percent energy extraction, and then the government offered financial support. Soon a wide new wave tank (the largest one in all of Europe) was built in the university, and work began in earnest. The Salter team began refining single units and devising ways to join the units together on a flexible spine.

At the first outdoor trial, in Draycote Water, a string of ducks built to one-fiftieth of full scale performed so well that the team immediately planned larger models, one-tenth of full scale, for a trial in Loch Ness. After that, the timetable calls for full-size assemblies anchored around the coast away from the shipping lanes.

Leaning against the university's sloshing wave tank, with the absorbed look that is characteristic of him, Salter is quick to describe his projected full-scale models. These will consist of giant steel and cement pods, each the size of a house. And each of these pods will pivot on a fixed horizontal spine. With pencil in hand, he likes to sketch what would happen, in the open sea, when the waves strike the ducks. The waves would activate the rotary hydraulic pumps. (See the diagram on page 54.) The pumps would pressurize the wa-

Salter's nodding duck (Harwell Reprographics Service)

ter, which would turn a turbine generator and produce electricity that would be transmitted to the mainland by seabed cable.

How much electricity are we talking about? The team believes that the average yield from the spines of full-size nodding ducks would be between 30 and 50 kilowatts per meter of seafront. A 300-mile (480-kilometer) chain of these ducks would, as mentioned earlier, be able to supply all of Great Britain's present electrical needs.

The target date? Long before 1990.

"And not to be overlooked," says Salter, "is the fact that the nodding ducks would seem to meet Mrs. Salter's requirements They are clean and safe, would produce large amounts of electricity, and will last a long time if not exactly forever."

Between the southern coast of England and the Isle of Wight flows a furious rush of water called the Solent. On its churning waves, in April of 1978, Sir Christopher Cockerell, Director of Wavepower Ltd., floated his Cockerell raft on its first sea trial. Riding the raft with him was his project manager and Mr. Alex Eadie, Parliamentary Under Secretary of Energy. According to Sir Christopher, trials of one-fiftieth scale had already been conducted. "But those trials were in wave tanks. Our purpose, today, on this one-tenth model is to gain at-sea experience." And the project manager added, "Today's sea trials will show people that wave power is not just a boffin's pipedream, but a tangible, credible proposition." He went on to explain that this was a fair trial because when the wind is from the southwest, the waters of the Solent develop waves that are

The Cockerell raft. On board are Sir Christopher Cockerell (*left*), Mr. Alex Eadie, Parliamentary Under Secretary of Energy (*center*), and Mr. J. Platts, project engineer (*right*). (Wavepower Limited)

just about one-tenth as ferocious and energetic as the North Atlantic waves off the coast of the Hebrides.

The raft was invented by Sir Christopher in 1971. It consists of three wave-contoured pontoons that are hinged together. As the waves rise and fall, so does the raft. This up-and-down motion compresses and extends hydraulic rams that connect the pontoons. These compressions and extensions pressurize a working fluid, which drives an hydraulic generator that then produces electricity.

While the raft flexes with the waves, computer-controlled recording equipment monitors its performance. This equipment, housed on a nearby barge, is connected to the raft by cable. There continuous printouts tell the story about wave conditions, power output, and stress.

Although the one-tenth-scale model produces only 1 kilowatt of electricity, Wavepower Ltd. calculates that a single raft 50 meters wide and 100 meters long (the length of a city block) might well under suitable sea conditions generate 2 megawatts. (A megawatt is equivalent to 1,000 kilowatts.) They hope to float a string of rafts soon in the open sea off the coast of Scotland and in the English Channel. There, under optimum conditions, a 15-mile (24-kilometer) string of rafts would be expected to add up to a 500-megawatt power station.

In the meantime, a number of problems still have to be solved. Needed are improved float design, easier and safer mooring capabilities, greater durability, and more corrosion resistance.

Sir Christopher, speaking for the global exploitation of wave power, said, cautiously, "Development is still in the early stages. But with continuing government support, a prototype wave-power station might be in operation sometime in the 1980s."

Altogether different from Salter's duck (a university-based project) and Sir Christopher's raft (a private, commercial venture) is the oscillating water column, a wave machine being developed at Britain's National Engineering Laboratory (NEL). It is a device designed to compress air by means of wave action and thus produce electricity.

The Cockerell raft, one-tenth-scale Solent tests (Wavepower Limited)

The oscillating water column (OWC) is an idea borrowed from Japan's Commander Y. Masuda, who invented the floating breakwater. Masuda found that if the breakwater was built like an inverted box with slots cut into the top, the wave height inside the box was significantly lessened—flattened by the air that came in through the slots. He also found that air, lots of it, was constantly forced into the box and out again by the up-and-down motion of the waves.

At NEL, an engineering team is constructing a model of the oscillating water column converter based on another device—the air pressure ring buoy. This is a hollow ring, shaped like a doughnut, with slots or holes on the top. It floats on the water, and as it bobs up and down in response to wave motion, the air that rushes in through the holes or slots is compressed and spins an air-driven turbine.

Still in the early stages of development, the oscillating water column converter is a more modern and sophisticated device that looks like a floating box with no bottom. It answers the question: How is it possible to keep the turbine spinning in one direction when the air that comes in through the slots at the top oscillates

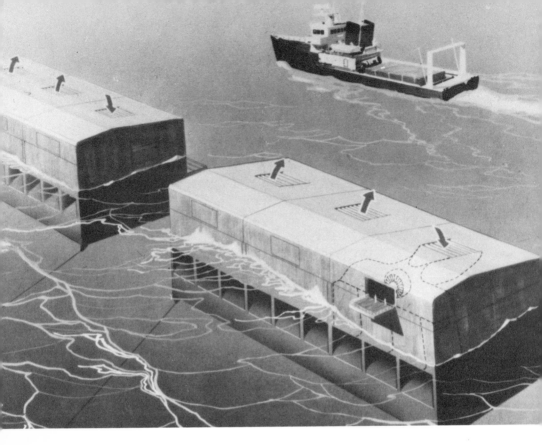

The oscillating water column converter (Harwell Reprographics Service)

(flows in two directions)—now in and now out? The answer is one-way flapper valves. These flapper valves direct the oscillating column of air in one direction so it can spin the turbine without interruption.

Also attracting attention is the Russell rectifier. This device rectifies (regulates) the motion of the waves so that the water operates a turbine only from one direction. (See the diagram on page 59.)

Here's how this regulator is set up. A series of boxlike reservoirs is anchored in the sea—some above the surface of the water, others below it. Placed between the high-level reservoirs and the low-level reservoirs is a turbogenerator. That's the structure, and here's how it works. The waves drive the water into the high-level reservoirs. The water drains down (through one-way flaps), activates the turbogenerator, produces electricity, and drains away. Highly favored by the National Engineering Laboratory, the Russell rectifier was still in the early stages of development in 1980.

Still another approach to energy extraction is under study in Israel. This is an integrated approach that would combine three sources of energy: wave, wind, and solar. None of these sources has ever been considered as *the* answer to the energy crisis because waves don't always crest at optimum height, winds don't always blow steadily, and the sun doesn't always shine with the same intensity for the same number of hours.

"Waves, winds, and sun are intermittent energy sources, but that's no reason for writing them off," says Professor Anthony Peranio, an American-born mechanical engineer serving at the Environmental Engineering Laboratories at Technion, Israel Institute of Technology in Haifa. "Instead, let's integrate them into one ener-

The Russell rectifier (Harwell Reprographics Service)

gy system. That way, we'll be able to develop a continuous supply of electricity."

Peranio has patented a number of innovative plans and models for extracting energy from alternate sources. One such is a combined wave-and-wind energy conversion device. As he explains it, this device could be rigged onto vessels that are connected to the shore by cable. It is a simple device, and it comes in two parts: a head tank mounted on the deck, and an inclined ramp or scoop that can be adjusted to varying wave height. (See the diagram below.) When the scoop is lowered into the sea, the waves rush up on it and spill over into the tank. Then the water is released to a low-head (low-pressure) turbine on the ship. This is no ordinary turbine. It can spin a generator to produce electricity, it can operate a compressor

Combined wave-and-wind energy converter (Anthony Peranio, Technion, Israel Institute of Technology)

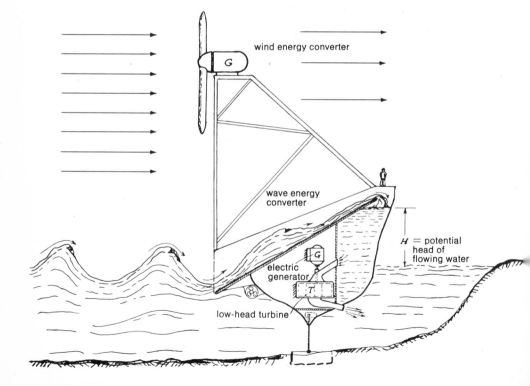

to produce compressed air, and it can turn a pump to lift water for storage and the subsequent production of electricity.

Simple as this converter is, it can generate sizable amounts of energy—more when the waves are ferocious, less when the sea is fairly calm. But even on a day when the waves rise to a height of only 6 to 7 feet (about 2 meters) and the wind runs at 16 or 18 miles (25 to 28 kilometers) an hour, it can generate a hefty 1.5 megawatts for every 325 feet (100 cubic meters) of scooped-up seawater. This output can be boosted by another 0.25 megawatts by catching the wind. All that is required is the installation on the ship of two high-speed propellers hooked in with the wave device.

The wave-and-wind converter might also be rigged on a series of offshore steel hulls that are moored in the bay as an energy-producing breakwater. According to calculations, fifteen such hulls, each 325 feet (100 meters) long, could supply most of the power needed by a community of 20,000 people. And this would be a fairly continuous power supply because the scoops would automatically adjust to the height of the prevailing wave conditions, and the hulls could be turned, oriented, and faced into the waves and wind as needed.

Peranio observes that if the wave-and-wind converter were integrated with Israel's solar power system, a really steady and uninterrupted power supply would result. "When the wind drops to a whisper," he says, "it's likely to be a clear and sunny day (good for solar energy extraction). On the other hand, when skies are cloudy, strong winds tend to prevail." He also points out that wave-and-wind converters need not be restricted geographically to Israeli waters. "They can be used to complement solar energy systems in the middle latitudes because there, when solar energy is most intense (in the summer), wind and wave energy is lax, and in winter, when solar intensity is low, wind and wave energy is greatest."

Israeli scientists expect such an integrated energy extraction system to work. The technology is simple, inexpensive, and available. Required capital investment is moderate, waves and winds are free, and additional jobs would open up for more people who will be needed to work on the construction and maintenance of these converters. But, as in other countries, the crying need is for money. The

Technion needs more government assistance to develop working models, at-sea tests, and pilot plants.

Actually, the Japanese were first with wave-energy converters that were practicable, durable, and efficient. These were the harbor buoys. They were designed by Masuda and powered by the waves as signal lights and harbor buoys. These buoys are of two kinds. Some are operated by a plunger-like mechanism that is activated by the vertical motion of the waves. Others are operated by a pendulum-like device that rocks with the rocking motion of the waves. Today, hundreds of these buoys are moored in Japanese waters and in waters around the world. They require very little attention and are easy to service.

The newest Japanese project is intended to tap Japan's enormous wave potential on a large scale. This is a $3.1 million floating experiment station. It is the 500-ton (450,000-kilogram) vessel, *Kaimei,* which was designed to operate in the Sea of Japan, where the waves average 10 to 13 feet (3 to 4 meters). *Kaimei* is kept afloat by four closed air compartments, and it's large—264 feet by 40 feet (80 by 12 meters). And it has twenty-two bottomless air chambers that gather wave energy. Here, the motion of the waves pressurizes the air which spins an air turbine that powers three generators.

Early *Kaimei* tests are encouraging, and further plans call for additional generators. After that, the goal is the construction of a 20-megawatt station that will send current to the mainland by underwater cables.

On a lesser scale, other nations are also researching wave energy extracting schemes. In the United States, at the Scripps Institution of Oceanography, Professor John Isaacs and engineer David Castel have been working on an ingenious scheme—a wave pump that amplifies the height of the waves as much as 13 feet (4 meters). It's a simple design and consists of a vertical pipe, a buoyant float, and a flapper valve at the bottom. So far, sea trials with a 300-foot (90-meter) hose have been successfully carried out in Kaneohe Bay, Hawaii. (See the photograph on page 21.) Plans call for pumping the water into a tank and building up pressure. The pressurized water would then be used to turn a turbogenerator and produce electricity.

At the University of Washington, an ocean-engineering seminar produced a number of feasibility studies: a Salter cam-type wave absorber, a fully sealed wave-activated buoy system, and an oscillating wedge scheme.

In Canada, however, activity on the wave energy extraction front is largely based on the British projects already discussed here: the nodding duck, the contoured raft, the wave rectifier, and the oscillating water column.

It's easy to see that these wave machines, although highly sophisticated, would only deliver considerable amounts of energy when wave action is energetic. For this reason they are mostly being tested (for future installation) in the stormy waters around the British Isles and Japan. As for environmental effects, about the only negative one that comes to mind is the inconvenience that might be caused to ships and fishing smacks by the floating mechanisms. But charted navigation lanes could take care of that problem. On the positive side, however, are two definite pluses. The floating mechanisms could serve both to attract fish and to decrease coastal erosion.

At this time who is to say which one of these energy schemes will turn out to be the most efficient and economical, the most durable and the most easily serviced? Also, who is to say which one will best handle sea conditions: the problems not only of unpredictable waves, but also of the forces exerted by winds and currents and tides? Will it be the floats or the ramps, the nodding ducks, the rectifier, or the oscillating water column? In the final analysis, will it perhaps be an integrated energy-extraction scheme that combines wave energy with wind energy and solar energy? Or will it be a scheme based on materials and designs yet to be invented? Only time will tell, and according to the experts, the time is not far distant. The target date? Most likely between 1985 and 1990.

In the meantime, let's consider another group of ocean scientists concentrating on yet another source of energy—the great currents that run like rivers through the sea.

5

Electric Current from

From the North Pole to the South Pole and right around the globe, water currents crisscross the ocean. There are many surface currents, long familiar to mariners. Here one flows east, there another flows west. Still others flow north or south, or around and around in great circles. Beneath these surface currents are others, often streaming in opposite directions; and below these countercurrents are still others, at many different levels, some racing one way, some another, and a few moving only sluggishly.

These currents spell energy—vast amounts of kinetic energy. So, in the last quarter of the twentieth century, some forward-looking ocean scientists proposed tapping the energy in the fast-flowing currents and converting it to electricity. Proposals led to plans, and plans led to studies and designs. Designs, however, need to be based on comprehensive information, and it has taken a long time to collect the necessary data about ocean currents: where they are, how fast they run, how steady they are, and what their temperature might be at different times of the day or month.

he Ocean Current

It was Columbus, while he was enjoying a speedy passage to San Salvador in a brisk, westward flowing stream, who made the first known mention of ocean currents. "I regard it as proved," he wrote in his *Journal*, "that the waters of the sea move from east to west as do the heavens." Since then, however, other mariners have observed other currents flowing in other directions.

In 1513, the Spanish explorer Ponce de León, who was searching for the Fountain of Youth, chanced upon a mystifying phenomenon. He was sailing southward off the coast of Florida in a furious storm. With the wild winds blowing in the direction in which he wanted to sail, and his ship seemingly making progress, he happened to look toward the shore and noticed that the ship was actually moving *backward*. Ocean demons? Evil spirits? No. Ponce de León didn't know it, but his little vessel was caught in the powerful current which is now called the Gulf Stream and which, in that part of the Atlantic, flows north. Although the winds blowing him south were strong, the northbound current was stronger.

Years later, in the interest of speeding ocean transport, Benjamin Franklin mapped this current as accurately as was possible for those times. He did this because he was challenged by a question he couldn't answer. "Why," he was asked on one of his voyages to England, "is it that British ships making for New York and points south

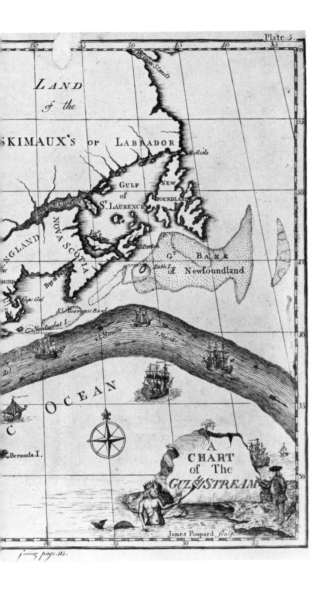

Benjamin Franklin's
map of the Gulf
Stream (Library
of Congress)

take two weeks longer to get there than the ships that sail for Rhode Island?" Since he didn't know, Franklin asked his cousin Timothy Folger, who was a whaler by trade. Folger knew what every whaler knew—that there was a current in that part of the ocean, a warm, fast-flowing eastbound current that the whales favored. "But the

English captains," Cousin Timothy explained, "follow their own charts, which make no mention of the current. So they get right into it, head-on, and buck it all the way across." He also said that the American ships made better time when they sailed eastward to Europe because they sailed *with* the current. Coming home, on the westward passage, they made sure to avoid it. Then Folger sketched that current for Franklin as he knew it to be.

After the Revolutionary War, Franklin, guided by Folger's sketch, plotted a more scientific map of the Gulf Stream. This he did by taking the temperature of the current on his many voyages across the Atlantic. Bundled in heavy winter clothing, with fur cap and boots, he prowled the decks, lowering empty bottles into the ocean, fishing up seawater, and checking it with a thermometer. In this way he discovered that the water of the current was invariably warmer than that of the surrounding water and of the air above it.

Today, oceanographers know that ocean currents differ from their surrounding waters in many ways. They differ in temperature, in speed, and in salinity. They also differ in kinetic energy as they run like "rivers in the sea" through all the oceans of the world. They are impelled by the prevailing winds, the rotational spin of the earth, and the varying temperatures and densities of the waters. But of all the surface currents, the Gulf Stream is probably the most important. Of this current, Matthew Maury, American naval officer, wrote in his *Physical Geography of the Sea* in 1855:

> There is a river in the ocean. In the severest droughts it never fails, and in the mightiest floods it never overflows. Its banks and its bottom are of cold water, while its current is warm. It takes its rise in the Gulf of Mexico, and it empties in the Arctic Sea. This mighty river is the Gulf Stream. There is in this world no other such majestic flow of waters. Its current is more rapid than the Mississippi or the Amazon, and its volume more than a thousand times greater. . . . Its waters, as far out from the Gulf as the Carolina coasts, are of indigo blue. They are so distinctly marked that their line of junction with the common seawater can be traced by the eye. Often, one half of the vessel may be

perceived floating in the Gulf Stream, while the other half is in the common water of the sea—so sharp is the line, and such the want of affinity between these waters and such, too, the reluctance, so to speak, on the part of the Gulf Stream to mingle with the littoral waters of the sea.

Maury's description of the Gulf Stream is now classic, and the vast amount of energy that is carried by this current is now recognized. But research on ocean currents was limited even in the nineteenth century, being prompted mainly by navigational interest and personal scientific curiosity. Data, accordingly, were accumulated slowly and painstakingly from entries in the logs of ocean-going ships.

Future data were to be collected by ocean scientists employed by the United States Navy, the Coast Guard, the United States Coast and Geodetic Survey, and private foundations. With more modern measuring instruments, they were able to draw up charts and tables

The major surface currents of the world's oceans
(Jerome Williams, *Oceanography*, Little Brown and Company)

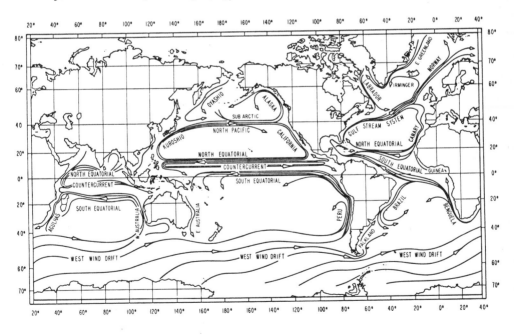

that showed the temperature and width and location of the Gulf Stream at various times. Recording these facts was particularly tricky because the Gulf Stream has a way of meandering; some say according to the phases of the moon. But in 1884 Admiral John Elliot Pillsbury was able to solve many of these mysteries. During his 1,000-hour tour of duty in the Florida Straits, he anchored his ship, the *Blake,* now in one spot and then in another and took readings with the latest scientific apparatus. His equipment included one special invention, his own current meter, which could simultaneously record the speed and the direction of the current.

Since Pillsbury's time, continued but inconclusive research has been conducted in all the oceans of the world. Precision instruments carried on many ocean-going vessels automatically record speed, temperature, location, and chemical composition of these rivers in the sea. In 1969, a superbly equipped government research submarine, the *Benjamin Franklin,* added considerable data to the growing ocean-current library. Two years later, Walter Duing, a professor at the Rosenstiel School of Marine and Atmospheric Science in Miami, Florida, added to that library. In a landmark study, he analyzed the greatest series of data ever obtained from the Florida Current, which is a portion of the Gulf Stream. Now for the first time, real data was available on "the temporal and spatial variations" of this current.

In the 1970s, the research focus changed. Because fossil fuels had become more costly, the emphasis, in part, shifted to the energy in the ocean currents: its availability in those running waters, its abundance, and its possible conversion to low-cost electricity.

It is generally believed that extracting energy from ocean currents would be a beneficial venture. It wouldn't pollute the environment to any great extent. It wouldn't require much in the way of energy inputs. And, since these currents run twenty-four hours a day, they would provide a continuously renewable source of energy. Such a venture could also be expected to be profitable provided the installations were large, durable, efficient, and cost competitive.

The first boost for the conversion of ocean current energy to electrical energy started quietly. Three friends, all ocean scientists, were chatting over a drink on a warm and sunny day and talking about Walter Duing's monumental work.

Time? It was 1973.

Place? It was the office of Dr. Harris B. Stewart, Jr., director of the National Oceanic and Atmospheric Administration's Atlantic Oceanographic and Meteorological Laboratories in Miami.

From the office windows they could see the Gulf Stream churning along—a bright blue current in the green Atlantic Ocean. As Dr. Stewart later told editor Arthur Fisher in an interview for *Popular Science* magazine, the three friends were, at first, idly admiring the blueness of the Gulf Stream. They knew that this current which runs in the Florida Straits between Miami and Bimini, carries more than fifty times the total flow of all the fresh-water rivers in the world. "We also knew," Dr. Stewart told Mr. Fisher, "that it ran steadily day and night at a good clip—4 knots. Then it occurred to us that we should investigate this Florida Current as a pretty good potential source of energy. . . . We figured that if we could somehow trap as little as 4 percent of this flow, we could extract somewhere between 1,000 and 2,000 megawatts of power. That's about the same as the output of the local nuclear power plant, so it seemed worthwhile."

Shortly afterward, Stewart and his colleagues met again and did a little figuring. They outlined a possible power plant for the Florida Straits. They even envisioned large propeller-driven underwater energy converters. The newspapers got word of this meeting, and the story of "underwater windmills" hit the front pages.

A Chicago millionaire, John B. MacArthur, also picked up the story. To MacArthur, a hard-headed businessman, the idea of inexhaustible, pollution-free energy in Miami's front yard made sense. He flew down to Miami. He talked with Dr. Stewart. He offered to underwrite a workshop for people with know-how: ocean engineers, heavy marine equipment specialists, turbine design engineers, energy economists, and oceanographers, plus corrosion and biofouling experts who deal with chemical and marine incrustations.

When the MacArthur Workshop met in 1974, the team was hand-picked. It represented money, professional expertise, and resource-fulness. There were delegates from the Florida Light and Power Company, Westinghouse, Allis-Chalmers, and the La Que Corrosion Laboratories of International Nickel. There were ocean engineers from three prestigious universities: Rhode Island, Massachusetts, and the University of California at Berkeley. There were interested parties from private industry and the government. After three days of discussion and papers and arguments and counterarguments, the delegates concluded that the energy in the Florida Current was equivalent to twenty-five 1,000 megawatt plants and that this current could definitely be tapped for the generation of electricity.

Attending environmentalists, however, questioned the unlimited tapping of this energy. As one put it, "Removing more than 2,000 megawatts from the Straits might cool the current and affect the climate of the eastern United States and northern Europe."

On the other side of the table, practical engineers considered such extraction energy devices as two-bladed propellors (the original so-called underwater windmills), rotors with cupped blades, and a system of giant water turbines. But the scheme that attracted the most attention was an underwater parachute system invented by Gary Steelman, who hails from land-locked Iowa.

A cutaway side view of the WLVEC—water low velocity energy converter—developed by Gary Steelman as an underwater parachute system (Proceedings of the MacArthur Workshop, National Oceanic and Atmospheric Administration, 1974)

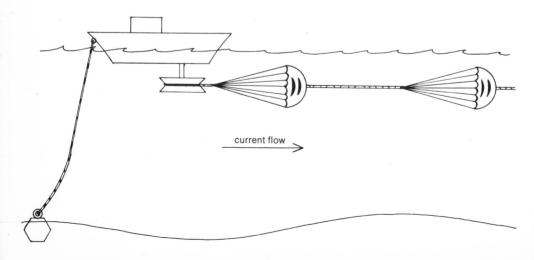

current flow

Steelman, a farmer with no formal training in engineering, had first used parachutes to trap energy from a small stream that flows through his farm. Since it worked there, he came up with the idea of using large parachutes to trap energy from the Florida current. His invention is as simple and inexpensive as it is ingenious. It is designed to convert the low-level flow of the current into usable power and is called WLVEC, "water low-velocity energy converter." The WLVEC consists of two components: one is a wheel on a shaft, mounted on a ship or a platform; the other is a continuous loop that runs like a conveyor belt around the wheel. Placed along this loop are sails that look like parachutes. These parachutes are constructed so as to expand as they face the oncoming current, then collapse as they round the loop and face away from the oncoming current. When this device is placed in the water, the parachutes automatically pull the loop along, and the loop is kept moving by the force of the current. Then, since the loop is trained around the wheel, it turns the wheel. And the turning wheel drives a turbogenerator that produces electricity.

The MacArthur Workshop ended on an optimistic note—that by the mid 1980s, substantial amounts of electricity from the Florida

A top view of the WLVEC (Proceedings of the MacArthur Workshop, National Oceanic and Atmospheric Administration, 1974)

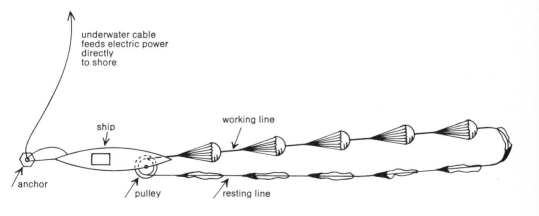

Current might very possibly be extracted at attractive prices. This energy might be available, then, in three forms: as electricity that would be transmitted to the shore by underwater cable; as hydrogen (extracted from seawater) that could be piped to the mainland as a gas or shipped there in cryogenic tanks; or as compressed air stored in "underwater balloons."

Before the workshop adjourned, the members recommended a research and development program. Shortly thereafter, a feasibility study found that the velocity of the ocean current varies from time to time and place to place, and this velocity is generally low. The energy level, therefore, is also low.

Low-level energy? For resourceful engineers, this was no insurmountable problem. They knew there was abundant energy in the ocean current and that it could be extracted with very large low-speed water turbines. And so a number of possible turbines were considered. The one, found the most efficient at that time was the vertical axis turbine (VAT). It had two pluses going for it. It could match the energy flow rate, and it could be adjusted to changes in flow direction. By 1980, however, Steelman reported that his new WLVEC model with 5-foot (1½-meter) parachutes worked perfectly. "It is designed," he said, "for a 10 horsepower output at sustained operation in a 2-knot current." According to his latest estimate, "There's no upper limit as to how much horsepower a WLVEC could produce. It depends only on the strength of the ropes and the materials used. With oil at $30 a barrel, an offshore Miami WLVEC could produce 1.5 million dollars of power a day."

When we talk about ocean-current power plants, we mean big money and government funds. Only one important federal grant had been approved in the United States in 1979. Under this grant, Dr. Peter B. S. Lissaman, vice president of AeroVironment Inc., in Pasadena, California, is trying to pick up where the MacArthur Workshop and the feasibility study had left off. He is trying to design an energy system around so-called underwater windmills—ocean turbines that were invented by two engineers: David Thompson and William Mouton.

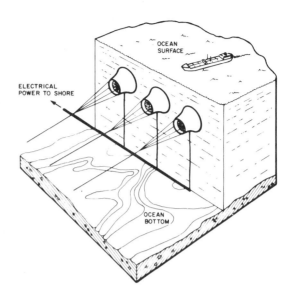

An artist's sketch of
AeroVironment's ocean
turbine energy system
(Mark E. McCandlish,
AeroVironment Inc.)

"Now imagine," says Lissaman, "a huge ducted underwater turbine, the size of two square city blocks, each with two giant fans in the center. And think of 250 of these turbines floated out to the Gulf Stream, sunk 75 feet beneath the surface, and anchored to the seabed with cables almost two miles long." He explains that "the speedy current will turn the fans. The fans will spin the turbines and produce electricity. And that electricity, transmitted to the Florida grids by underwater cable, could easily supply a good portion of the state's electrical needs."

This project, this ocean turbine energy system, is called the Coriolis Program in honor of a nineteenth-century Frenchman, Gaspard de Coriolis. A mathematician and engineer, he is remembered for the Coriolis theory, which describes the movements of the ocean currents and air currents in response to the earth's rotational spin. According to this theory, the currents in the Northern Hemisphere are deflected to the right, or clockwise; in the Southern Hemisphere, they are deflected to the left, or counterclockwise.

Robert Radkey, manager of the Coriolis Program, describes the system with great enthusiasm. "We expect few problems concerning the mounting and mooring of our turbines and generators. That's because they'll be housed in the ducts, which look like giant coffee cans with both ends removed. In these ducts, stability is increased. And power extraction by the rotors is also augmented. As for biofouling, we expect very little of that because marine organisms rarely attach themselves to things in running water. And the Gulf Stream certainly runs—4 knots an hour."

Asked whether the speed of the Gulf Stream varies with the seasons and if so, what effect it would have on electrical generation, Radkey said, "We know there's a difference in speed between summer and winter. We've studied that and, at this early stage in the program, we estimate a rated capacity for Coriolis of 57 percent. And that's not bad at all."

The problem raised by some environmentalists—that the extraction of heat from the Gulf Stream might affect the climate of the eastern United States and northern Europe—is brushed aside by both Lissaman and Radkey "because Coriolis doesn't extract heat. Our turbines simply *intercept* the kinetic energy in the running water."

Any problems of slowdown of the current? "Well, when the turbines intercept the current, they may slow it down as much as 1 percent. And though we don't think of a 1 percent slowdown as significant, we will continue to study it."

Prototype unit of Coriolis I on tow to the Gulf Stream mooring station (Mark E. McCandlish, AeroVironment Inc.)

Based on their research, spokesmen for AeroVironment believe that Coriolis will be capable of producing 10,000 megawatts without pollution. As for the effects of all that turbine activity in the ocean, "Our studies show that in the Florida Straits local wave-making is below the normal [wind-induced] sea state." They also claim that any current-speed and temperature perturbations caused by Coriolis activity would be below natural fluctuations.

In the meantime, more studies are needed. Needed too is the construction and testing of a 39-foot (12-meter) ocean turbine model. After that, industrial backing could have the full-scale Coriolis operating on-station, within thirty-six months of start-up of construction.

We are told that the ocean currents run with enough energy to electrify the world. Is this then the answer to the energy shortage?

In a global sense, it's too soon to evaluate the future potential of this energy source. And most of those organizations or individuals who would be in a position to invest money in such a project take the approach that as yet not enough is known about the currents that crisscross the ocean: their speed and temperature, their steadiness and energy level.

Dr. Harris B. Stewart, Jr., the organizer of the MacArthur Workshop, thinks otherwise. In a telephone conversation with this author, he said, "It isn't necessary to wait till all the facts are in for all the currents in the world. Let's start with the Florida Straits. There's lots of energy there. We know that, and we know it's available for extraction. Let's not concentrate solely on any one source. What we have to do is utilize the energy sources that are available *locally*. In California, we extract geothermal energy. In Arizona, we extract solar energy. In Cobscook Bay, we extract tidal energy. In the North Atlantic, off the Scottish coast, we extract wave energy. And in the Florida Straits, we extract ocean current energy."

Still speaking quietly, but with conviction, Stewart added, "What we must *not* do is what the United States Department of Energy is doing right now—putting most of its solar energy eggs into one energy basket and going all out for the development of one energy source—OTEC—ocean thermal energy conversion."

6

OTEC–Ocean Therma

In his office at Newport Beach, California, Harold Ramsden of Global Marine Development Inc. shook his head at the mention of Gulf Stream energy, wave energy, and tidal energy. "The real energy bonanza," he said, "is in OTEC: ocean thermal energy conversion. Sounds formidable? Let me put it this way. Ocean thermal energy is nothing more than solar energy that's been absorbed by the ocean. But note—there's also very cold water in that ocean, about half a mile below. This gives us a temperature difference. With twentieth-century technology, we can take advantage of this temperature difference, extract energy from it, and convert that energy to electricity. And there you have it, ocean thermal energy conversion—OTEC."

He flipped open a 1978 government report and found the statement of James Madewell, assistant director of the United States Department of Energy (DOE) that OTEC can now supply baseload power at prices competitive with coal and nuclear power. "But we have to remember," said Ramsden, "that we have some way to go

nergy Conversion

yet to be sure of our economics. The success of OTEC begins with the temperature difference in the ocean, and this difference is not as favorable everywhere as it is around Hawaii, Puerto Rico, and some of the other tropical islands."

When ocean scientists speak of temperature difference, they often call it "delta T" and write it as Δ T. Ocean engineers study delta T, which varies from place to place and design OTEC power plants. Naval architects then study the design and plan the housing for it, in huge, specially constructed OTEC ships or platforms.

Ramsden was standing near a bulletin board, which he calls his "brag board." It was covered with diplomas in naval architecture and marine engineering, as well as citations in oceanographic research. He pointed at a picture. (See the photograph on page 80.) "It's the artist's rendering of OTEC 1. That's our 1-megawatt test ship, and it's sponsored by DOE. It will be moored off the coast of Hawaii to test full-scale equipment. Ultimately, we'll be generating electricity for commercial use."

OTEC 1—the retrofitted U.S. naval vessel *Chapachet*
(Global Marine Development Inc.)

The artist's conceptualized picture of OTEC 1 is impressive. It shows the converted United States naval tanker *Chapachet* being retrofitted as a home-away-from-home for the crew of thirty-one men and women who will be handling the job of ocean thermal energy conversion. Plans call for comfortable quarters for scientists, environmentalists, test engineers, computer analysts, deck hands, and cook. A van, to be installed on deck, will house the data processing and computer control center, and that indispensable machine, the ice-cream maker, will be within snacking distance. Look at the cutaway section (see the above photograph) for the heat exchanger, the pumps, and the long cold-water pipe that will bring the water up from the depths to cool the vaporized ammonia in the condenser.

"But that's not the whole story," says Ramsden. "In the next three to five years, we'll be installing and tearing out any number of heat exchangers, pumps and pipes for comparison testing. And we'll be working out ways to improve the electric generators. We'll also be researching the problems of corrosion and biofouling."

As any marine engineer will tell you, corrosion is one of the big headaches in this business of producing electricity at sea. Because salt water corrodes (eats away the metal components it comes in contact with), the life of the power plant is naturally limited. The other big headache is biofouling—clogging of certain components with tiny ocean plants and animals. This clogging retards the transfer of heat and reduces the effectiveness of the heat exchangers.

The test site for OTEC 1 was carefully chosen. It is only 18 nautical miles (21 miles) northwest of Keahole Point and near Kawaehae. This is a deep-water Hawaiian port where back-up service is available. Here the wind, current, and wave conditions are mild enough so the ship can be kept "on station" by a line from the bow to the mooring buoy. Should unusual storms pass by, the ship is prepared. It will stay on station with electrically driven steerable thrusters.

The idea of using the ocean's thermal difference to develop electricity surfaced only about a hundred years ago. The time was 1881. The occasion was the publication of a paper on solar sea power by Jacques D'Arsonval, a French physicist.

At that time, engineers and ocean scientists already knew a great deal about the ocean's capacity to absorb and store solar energy. They also knew that the ocean is in thermal disequilibrium—that below the warm surface layer is a dense layer of cold water. At the same time that the surface waters warm up and start warm currents flowing toward the poles, the polar icecaps melt under the polar sun, slowly of course. This meltwater is cold and dense and heavy; consequently, it sinks and flows off in the only direction possible— toward the equator. This is an endless cycle, guaranteeing an endless supply of warm water underlain by cold water. As one oceanographer put it, "It's this circulation pattern of the ocean that explains the presence of the delta T, the temperature difference between the surface and the depths."

Although all this was known in 1881, almost fifty years were to pass between the publication of D'Arsonval's paper and the first attempt—by another Frenchman, Georges Claude—to generate electricity from the ocean. It took time for D'Arsonval's proposal to catch on because not all scientists and engineers were familiar with the laws of thermodynamics. Of particular importance was the law that postulates the flow of heat from a higher temperature source to a lower temperature source. (This is easy to see if you place an ice cube on a dish and hold your finger firmly on the cold cube for a few minutes. You'll feel a depression forming beneath your finger, a depression that's created by heat flowing from your warm finger to the cube.) When D'Arsonval considered this law of thermodynamics with reference to the Δ T in the ocean, he had but one thought. "*Violà!* If we should float a machine [a heat machine, a heat exchanger] in the ocean, I doubt not but that the flow of heat from the warm surface water to the colder depths could be converted to do useful work—to turn a turbine. And if we can turn a turbine, we're on our way to generating electricity."

D'Arsonval, who was ahead of his time, urged the use of ammonia rather than seawater as a working fluid. The use of ammonia, he argued, would cut down on the size of the turbines. But that other Frenchman, the very determined and stubborn Claude, who in 1929 actually built the world's first OTEC project, didn't agree.

Claude built his OTEC power plant on land at the edge of Matanzas Bay in Cuba. He announced he would produce electricity in an *open cycle system*—that is, he would use the warm seawater to run a steam engine, and he would then allow the used water to flow back to the sea.

Now, under normal atmospheric conditions, water boils at 212 degrees F (100 degrees C), but if you lower the atmospheric pressure, you lower the boiling point. So Claude designed a vacuum pump to reduce the pressure in the boiler and got seawater to boil at a lower temperature, at which point it flashed (turned) into low-pressure steam. This steam spun a low-pressure turbogenerator. And so Claude achieved what he had set out to do. He produced electricity by using seawater as the working fluid. He produced 22 kilowatts (22,000 watts), of electricity, enough to illuminate 220 one-hundred-watt light bulbs, but hardly enough to start a utility company. It was, nevertheless, a proven beginning.

He also got a fringe benefit in the form of fresh water, which is often in short supply in tropical regions. This is how it happened. After the steam had turned the turbogenerator, it flowed to the condenser. The condenser was continually cooled by the cold water brought up from the depths in a long pipe. In the cooled condenser, the steam (which contained no salt) condensed into fresh water. You wonder what happened to the salt? It was left behind in the boiler. (Remember that left-behind salt—it was to create a big headache for Claude because, as you know, salt corrodes metal.)

In answer to his detractors, who sneered and asked him, laughingly, how many kilowatts of electricity he had put *into* his system to get those 22 kilowatts out of it, Claude sent a formal statement to *Mechanical Engineering* in December 1930. "First of all," he wrote, "let me explain exactly what we have been trying to do. We have not been endeavoring to extract power from the waves, from the tides, or from the streams. What we had in mind, my friend Boucherot and myself, was to utilize the remarkable fact that, in tropical seas, through the paradoxical collaboration of the sun and the poles, an important and almost invariable difference of temperature is maintained between the surface seawater, which is continually heated by the sun to 75 or 85 degrees F, and the deep-sea water which, because of the very sluggish flow from the poles to the equator, does not rise much above the freezing point—that is, above 40 to 43 degrees F—at a depth of 3,000 feet."

Well, so far so good, except for that long cold-water pipe, which was the undoing of Claude's experiment. To reach the deep cold water, the pipe had to run at a slant from the installation, which was land-based. And to maintain control of it, Claude made sure that the pipe, which was an astonishing 1½ miles (2½ kilometers) long, was secured to the sea bottom—rigidly secured. A severe storm struck, and the pipe was wrecked, destroyed, and carried away. Then, since so much money would have been needed to replace it, the project was abandoned.

Still, the Matanzas experiment was not altogether a failure. It did produce 22 kilowatts of electricity long before the storm, and although this was a ridiculously small amount (especially when you think of the time and money that went into the project), it proved that it was possible to generate electricity from the ocean by taking advantage of its thermal disequilibrium.

One wrecked project could not discourage Claude, so he turned his attention to other OTEC plants, trying other angles. Off the coast of Brazil he tested a floating platform, the converted cargo ship *Tunisie*. This time he hoped to generate 800 kilowatts of electricity. Then he attempted a larger installation at Abidjan, on the Ivory Coast of Africa, an installation that was never completed. Everywhere he met with the same two problems: a great deal of corrosion created by the warm, salty seawater in the boilers, and huge expenditures required for the giant equipment.

Claude finally abandoned his OTEC ventures, but his country didn't. After World War II, the French government formed a state corporatation called Energie des Mers. This corporation built a 75-megawatt (75,000-kilowatt) sea thermal plant in the Mediterranean, shipped it piece by piece to the Gold Coast in Africa, put it into operation, and went broke.

But OTEC was too good a concept to be dismissed, passed over, and relegated to limbo with "Yes, good idea but impractical and unrealistic." The Western world had become energy dependent—had in fact, jumped on the energy merry-go-round. It was sending men to the moon and to the deeps of the sea, racing faster cars, driving larger tractors, raising taller skyscrapers, with ever more powerful backhoes, airhammers, and cranes. And the military! Worldwide, the military had mechanized its war machines and become the number one consumer of energy on planet Earth.

And so, men and women of vision became increasingly interested in new sources of energy, especially from the sea. The French pushed for the development of the world's first great hydroelectric tidal power plant on the Rance River. Canada and the United States conferred endlessly about a tidal installation in turbulent Passamaquoddy Bay. In the British Patent Office, inventors by the score filed plans for the extraction of energy from the waves that roil the global seas.

In the 1960s, once again OTEC proposals surfaced, this time conceived by the Andersons, a father-and-son engineering team, working out of York, Pennsylvania, with their own company, Sea Solar Power, Inc.

The Andersons are known for their calculations which show that OTEC installations could actually produce electricity at lower cost than coal or nuclear plants. Other engineers disagree with this opti-

An artist's conception of an OTEC platform (Sea Solar Power, Inc.)

mistic prediction. "We only wish it were so. We hope it is so. But we're still feeling our way." The Andersons, however, base their mathematics on their design concept. This design, first published in 1966, calls for a large floating platform, much like a present-day oil rig, at anchor. Topside, the platform provides quarters for the crew and for their diving gear and other supplies. It also provides space for keeping materiel: cranes, hoists, tools, and the like. The power plant itself is submerged. This, the Andersons claim, adds stability to the plant and equalizes the pressure of the working fluid *inside* the heat exchanger with the pressure of the water *outside*—an important factor minimizing heat-exchanger cost. Additionally, the condensers are located above the evaporators, thus eliminating the need for a feed pump.

The Andersons' proposed power plant resembles Claude's in some ways but differs significantly in other ways. While Claude used seawater, a free source of fuel, as the working fluid in his open-cycle system, the Andersons called for the use of liquid propane as a working fluid in a closed-cycle system, generally referred to as the Rankine cycle. They pointed out that with liquid propane the plant would be able to use a small high-pressure gas turbogenerator instead of the monster-sized low-pressure steam turbogenerator Claude had to use. Furthermore, a working fluid such as propane, Freon, or ammonia is non-corrosive.

In concept, the closed-cycle system is pretty simple. Pumps circulate the working fluid through the power plant, which is a mass of pressure vessels, piping, and machinery. First, the working fluid is pumped into the evaporator. Here, warmed by the circulating ocean water, it turns into a gas. The gas then flows to the turbogenerator, spins it, and produces electricity. However, this gas is not used up or expelled from the plant—not at all! After its work in the turbogenerator is done, the gas flows along to the condenser, where it condenses and liquifies. Then, as a liquid, the working fluid is again pumped to the evaporator and the cycle is repeated. It is broken only when the plant is closed down for periodic maintenance or repair.

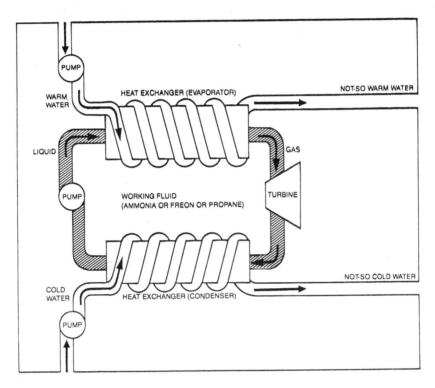

An OTEC closed-cycle system (*Renewable Ocean Energy Sources,*
Office of Technology Assessment, May 1978)

By the 1970s, a number of government-sponsored research teams
(especially in the United States) began investigating the OTEC concept for the production of electricity. Heading the list was Sea Solar
Power, Inc., along with two universities: Carnegie Mellon and the
University of Massachusetts. All three bypassed the open-cycle system and opted for the closed-cycle and the use of a liquid refriger-
ant as a working fluid. All, however, differed on questions of design
and construction materials.

As the calendar turned to the 1980s, two industrial organizations
funded by the National Science Foundation and/or the Department

of Energy pushed into the foreground. One is Lockheed Ocean Systems Division, the other TRW Ocean and Energy Systems.

Lockheed was working out the design for a vertical semi-submerged spar-type structure with a 250-foot (77-meter) central core. Except for the crew's quarters and maintenance facilities, the plant is submerged. This makes it stable and hurricane-proof, important assets for a structure destined to be moored in the sometimes hurricane-lashed waters around Hawaii, Florida, and the tropical islands in the Caribbean and the South Pacific.

In the Lockheed design, ammonia is used to generate electricity in several interchangeable power modules, each containing its own evaporator, condenser, turbogenerator, and pumps. These are attached to the outside of the core structure. (See the photograph on page 89.)

As for the platform, that is made of prestressed reinforced concrete, but the 1,500-foot (450-meter) cold-water pipe will probably be made of lighter and more flexible material. The entire plant, held in place by a single mooring line, is expected to produce 160 megawatts of power—enough to meet the needs of 160,000 people ashore. It could be in operation by 1986.

In the meantime, in June 1979 Lockheed's Mini-OTEC, the world's first closed-cycle self-sustaining OTEC system, began operating at sea. It has a rated capacity of 50 kilowatts (50,000 watts). It was assembled in a shipyard in Honolulu, housed on a barge, and moored nearby for the blessing and commissioning ceremonies on May 29.

Such ceremonies, such festivities! Bands played. Flags snapped in the breeze. Hawaiian dancing girls tossed orchid leis to the bystanders, as the television cameras ground away. And VIPs invited by the governor made rousing speeches about this global "first" that was ushering in a new era of day-and-night ocean energy production.

As Lockheed's C. Turner Joy, Jr., the manager of new business development for the company, summed up the situation, "A few more tests will prove beyond the shadow of a doubt that the time for OTEC has come. Even the doubting Thomases are just about ready

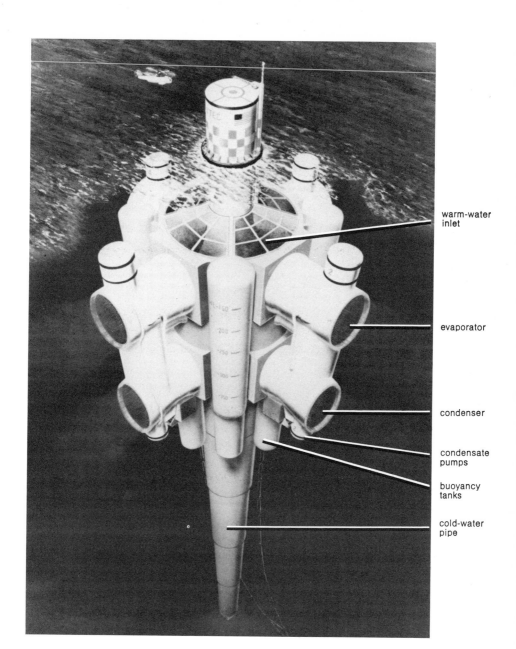

warm-water
inlet

evaporator

condenser

condensate
pumps

buoyancy
tanks

cold-water
pipe

An artist's conception of an OTEC platform
(Lockheed Missiles and Space Co. Inc.)

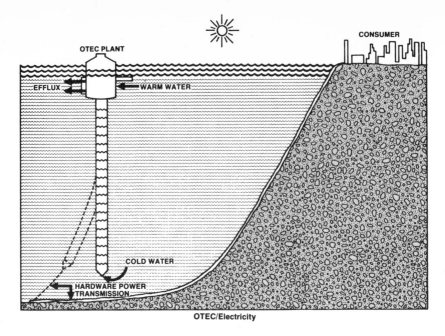

OTEC PLANT

CONSUMER

EFFLUX

WARM WATER

COLD WATER

HARDWARE POWER
TRANSMISSION

OTEC/Electricity

A simplified schematic of an OTEC power plant (*Renewable Ocean Energy Sources*, Office of Technology Assessment, May 1978)

to agree that the system has been studied long enough and it's time—as we say in the aerospace industry—'to get on with the tin bending.'" He further predicted that Mini-OTEC will be producing 50,000 kilowatts in the near future.

Racing with Lockheed to develop and operate a moored OTEC system is TRW. TRW "hardware" (the heat exchangers, pumps, and cold-water pipe) will be installed in OTEC 1—the old naval tanker that Global Marine is retrofitting under DOE sponsorship. As of 1979, this project was planned to be a simulation exercise, not yet a turbo-driven system. But according to Raleigh Guynes, spokesman for TRW, "Within a decade, you'll be seeing not one but a whole

Mini ocean thermal energy conversion (OTEC) system commissioned May 1979 off Honolulu (Lockheed Missiles and Space Co. Inc.)

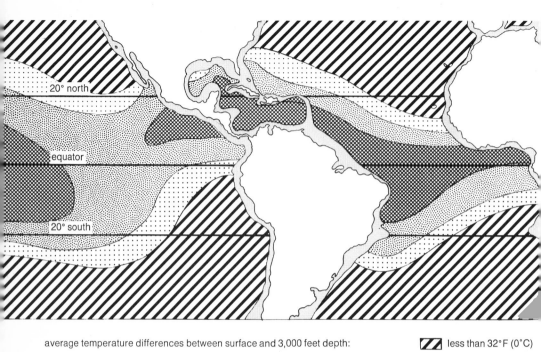

average temperature differences between surface and 3,000 feet depth: ▨ less than 32°F (0°C)

A thermal resource map of the world's oceans (*Technology Review* from

fleet of floating OTEC plants, converting the ocean's thermal energy into thousands of megawatts of electricity."

Guynes is eager to show off TRW's 6-foot (2-meter) OTEC model, which is displayed in the lobby of the company's Redondo Beach (California) headquarters. With unabashed confidence, he points to the model's special features. Only a few of these features are now being incorporated in OTEC 1, but all will be included in the redesigned circular platform. Then, perhaps within three to five years, the typical TRW platform will have a deck as big as two football fields. It will be seventeen stories high. Ammonia will be used as the working fluid, and the plant will receive its cold water from 2,100 feet (62 meters) below, through a bundle of three intake pipes. Each of these pipes will be 48 inches (1¼ meters) in diameter and made of reinforced polyethylene. These pipes are virtually standard items. They are flexible, noncorrosive, and resistant to biofouling,

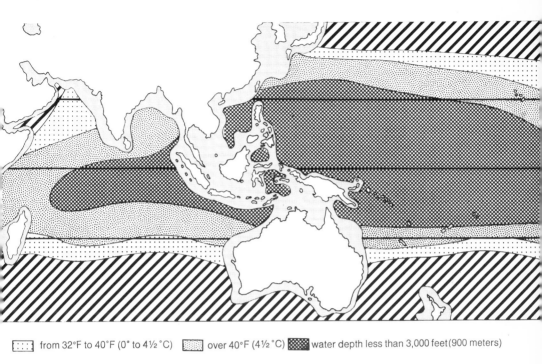

[::::] from 32°F to 40°F (0° to 4½ °C)   [::::] over 40°F (4½ °C)   [▓▓▓] water depth less than 3,000 feet (900 meters)

the U.S. Department of Energy)

and they bear up well to the hydrostatic pressure of the ocean because that is balanced by the pressure of the water within the pipes. All these are important cost factors, not only initially, but in regard to maintenance.

Guynes sees one major problem, which stems *not* from a lack of technology but rather from a lack of a systems approach. "It doesn't help one bit if one subcontractor goes all out to build the smallest possible heat exchanger, another builds the most elaborate set of pumps, a third creates indestructible cold-water pipes, a fourth develops a super mooring device. The components must serve each other in size and function, in durability and effectiveness. Progress can best be made in terms of the whole system, which is designed to perform one ultimate function. This system needs to be designed and understood in advance. Too often, with government contracts, work is farmed out to many subcontractors, each one doing his own thing. It's mighty hard to put the whole project together and create an effective system after the components are gathered together."

There's yet another problem that developers often mention. That's the matter of polarization of attitude. This polarization keeps the big commercial money away from OTEC investment and retards progress. "For good reasons or bad," many developers say, "we are constantly harrassed by the utility companies, the banks, and the nuclear establishments."

Despite these institutional blockades, there's general agreement that if OTEC comes through with plenty of power at attractive prices, the public will buy it. Then fleets of OTEC ships, moored and anchored around some of the coasts, will indeed become a commonplace—but only in the warm ocean belt, which lies between 20 degrees north and 20 degrees south of the equator. It's there that the solar intensity is greatest. In addition, four favorable conditions must exist:

1. The ΔT, the temperature difference, must be at least between 35 and 40 degrees F (2 to 4 degrees C).

2. The cold-water layer should begin no deeper down than a scant kilometer (about ⅝ mile), or the cold-water pipe would be too costly to service.

3. The distance from shore must not be too great, or running underwater transmission lines would be uneconomical.

4. Ports of call with back-up repair service must be nearby.

Although the greatest ΔTs occur mostly in midocean, there are, nevertheless, offshore sites with enormous potential. (See the chart on the next page.) For the United States, mark down the Gulf of Mexico and the waters around Florida, Hawaii, Puerto Rico, Guam, the Virgin Islands, and Micronesia. According to the United States Office of Technology Assessment, OTEC exploitation of the 3 million square miles of Micronesian waters alone might solve our energy problems because the equivalent of 47 percent of our present American electricity production could be extracted there. But think of Micronesia's location and its distance from back-up ports and markets!

POTENTIAL OTEC SITES

showing minimum distances (in kilometers) from coast to site

Countries bordering the Indian Ocean
(clockwise order)

Madagascar	32
Mozambique	25
Tanzania	25
Kenya	25
Somalia	25
P.D.R. of Yemen	32
Oman	6
Iran	32
Pakistan	32
India:	
West coast	120
East coast	65
Burma	75

Countries bordering the Pacific Ocean
(clockwise order)

Hawaii	10
Mexico	25
Guatemala	32
El Salvador	65
Honduras	75
Nicaragua	95
Costa Rica	7
Panama	25
Colombia	25
Ecuador	25
Australia:	
Northeast corner	65
Otherwise	300
West Irian	5
Java	5
Philippines	5
Vietnam	75
Sumatra	50

Countries bordering the Atlantic Ocean
(clockwise order)

Sierra Leone	50
Liberia	50
Ivory Coast	50
Ghana	50
Benin	50
Cameroon	65
Brazil	
1° to 20° south	15
Otherwise	100
French Guiana	130
Suriname	130
Guyana	130
Venezuela	3
Colombia	32
Panama	25
Costa Rica	15
Nicaragua	150
Honduras	24
Mexico	7
United States of America:	
Florida	1
Puerto Rico	6
Cuba	2
Jamaica	2
Haiti	2
Dominican Republic	2
Guadeloupe	2
Dominica	5
Martinique	2
St. Lucia	2
St. Vincent	2
Grenada	2

Source: Lavi, A. "Plumbing the Ocean Depths: A New Source of Power," Institute of Electrical and Electronics Engineers *Spectrum*, 10, 22–27 October 1973.

In the Third World, good OTEC sites are fairly plentiful near countries with exploding populations and rising expectations. These countries need cheap electricity. "For such countries," says Dr. William B. Cutler, systems design specialist for Lockheed, "an OTEC plant is a fine form of foreign aid. It can be a small plant, say 10 megawatts (unlike coal or nuclear which go to 1,000 or more megawatts). And small plants are suitable for small developing countries, which can, at most, use only a few tens of megawatts. Further, OTEC technology is simple and benign."

But the size of an OTEC ship? And the mechanical problems?

"Aside from the large size of the equipment," Dr. Cutler goes on to say, "an OTEC power plant is very much like a refrigerator and the concepts are easy for a mechanic to grasp. Also, there's a minimum of danger to workers and neighboring inhabitants. And much of the repair and maintenance can be done locally."

When an attractive resource becomes tantalizingly economic, new secondary ideas are quick to surface. One such idea is the development of OTEC electricity not for ship-to-shore transmission, but rather for *open-ocean on-board factories*. With an on-board factory, electricity is first produced on large OTEC plant-ships and then used right there to power on-board manufacturing.

A large factory ship would not have to be anchored and moored near shore. Consequently, it would not need ship-to-shore power cables. Such a ship would be able to rove, cruise, or graze the high seas at a speed of about ½ knot (about ½ mile an hour) in search of the most favorable ocean temperatures because it's a fact that, at any given place, the Δ T is not absolutely steady throughout the year. Maintenance problems will be complicated, however, due to drag on that long cold-water pipe, and divers will have to be on call constantly.

Front-runner for developing an on-board factory ship (which some call an instant deep water port) is Dr. Evan Francis, head of the Applied Physics Laboratory (APL) at Johns Hopkins University. As Francis explains it, "The APL plant-ship, now on the drawing boards, will be built to produce ammonia at sea."

Why ammonia?

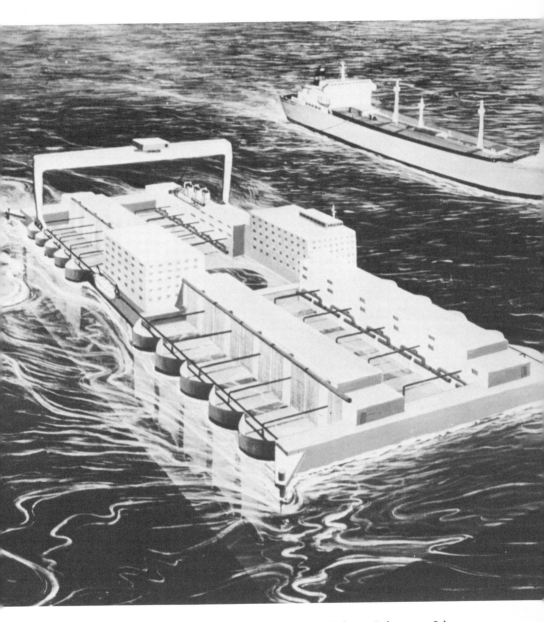

A roving OTEC factory ship designed by the Applied Physics Laboratory, Johns Hopkins University. The ship would house processing equipment and provide storage and accommodations for thirty-one persons. (U.S. Department of Energy)

"Because ammonia is the main component of fertilizer," says Dr. Francis. "Without ammonia, no fertilizer. No fertilizer, reduced food production and massive starvation as world population continues to increase."

Dr. Francis points out that the production of ammonia requires only hydrogen (from seawater) and nitrogen (from the air). Ammonia in liquid form could be shipped to distant shores in refrigerated tankers. There it could be used as a feedstock of fertilizer.

In the United States at present, ammonia is produced from natural gas. It's estimated that by 1985 as much as 7 percent of our natural gas consumption will be required for ammonia production. "But," says Francis, "by producing OTEC ammonia we will, in fact, be conserving 7 percent of our natural gas supply."

Furthermore, ammonia can be used in the production of synthetic gas, and it can be decomposed back to hydrogen and nitrogen, and hydrogen can be used in fuel cells to generate electricity. Additionally, the hydrogen can be stored in metal hydrides for reconversion to electricity.

Under consideration, too, is the installation on board these factory ships of special furnaces to smelt nickel and molybdenum. But what particularly interests the major companies is aluminum. A tremendous amount of aluminum is used in the Western world, and a tremendous amount of electrical power is required for its production. With OTEC electricity produced on-board, the factory plant-ship is in business. Bauxite, the principal source of aluminum and which is so plentiful in Jamaica, Brazil, Guinea, and Australia, could be loaded onto nearby plant-ships and manufactured into aluminum. The slag could then be dumped into the sea and the aluminum ingots boxed and shipped ashore.

Taking their cue from APL, Lockheed, TRW, and Sea Solar Power, Inc., other companies are also beginning to consider the development of OTEC plant-ships as ocean energy industrial complexes. But this is a distant vision. In the near future, roving OTEC plant-ships will probably be used to retrieve valuable products from the sea: liquid hydrogen, oxygen, chlorine, and methanol.

There's yet another plus to be considered in the matter of OTEC development, and that's the political factor. Grazing or roving the high seas, a plant-ship flying the flag of the United States (or any country) under the present law of the sea could neither be appropriated nor taxed by another power.

What about the effects of OTEC activity? Any adverse possibilities?

In the opinion of most scientists there may be some, but these won't be known for a while. There might be some dislocation of marine ecosystems locally, but this would probably be offset by the nutrient-rich water brought up by the cold-water pipe. There would probably be some toxicity from the metals of the heat exchangers and the anti-fouling agents used, as well as from working-fluid leaks. Additionally, there might be some oceanic effects from the transfer of heat from the surface to deeper water. There might even be a change in the climate locally if the carbon dioxide in the deep cold water is released to the atmosphere on a considerable scale. OTEC engineers, however, point out that it is most *unlikely* to happen because the cold water will be returned to the depths without ever being exposed to the atmosphere.

In the main, however, even the skeptics are beginning to see OTEC as an attractive source of energy that's available twenty-four hours a day, rain or shine, although it is limited to certain tropical waters. But then, no energy source (new or old) is available everywhere. What we need to do is develop each source where it works best. In this way, we'll be able to satisfy our energy needs.

7

Ocean Energy Farms

Energy buffs have a great deal to say about the conversion of ocean energy. Creatively, they would harness the waves and tides. Imaginatively, they would intercept the fast-flowing currents. Ingeniously, they would draw energy from the ocean's thermal disequilibrium.

All these schemes are conceivable and possible. All are novel and different. Yet, in one sense, they are all the same. All would extract energy from the ocean by placing a wheel in the water, getting the wheel to serve as a turbine, and getting the turbine to spin a generator, thus producing electricity.

It was inevitable that some bright inventor would think up an entirely different scheme for the extraction of energy from the ocean. That innovator was Dr. Howard A. Wilcox at the Naval Ocean Systems Center, first mentioned in Chapter 1.

"Here's what we have to do," says Wilcox. "We have to *grow* our own fuel. And there's plenty of space in the ocean for any number of energy farms."

What plants would he use? "Kelp, because kelp flourishes natural-ly in the coastal waters. It absorbs and stores tremendous amounts of solar energy, and it grows fast—an astonishing 1 to 2 feet ($\frac{1}{3}$ to $\frac{2}{3}$ meter) a day."

Kelp has long been known as a useful plant. In Japan, it has been for centuries a standard item on the menu. In the United States, it has been used as an emulsifier in salad oil and an ingredient in such items as plastics and paint, cardboard and toothpaste, and animal feed. More recently, kelp as a potential source of methane, which is a substitute for natural gas, has begun to intrigue marine scientists and even to attract the attention of some industrialists.

"We could transplant kelp to ocean energy farms," says Wilcox, "cultivate it, harvest it, and convert its stored energy to methane gas and ethanol for the tanks of cars and trucks, buses and planes."

Very shortly thereafter, Wilcox, who works at lightning speed, had plans drawn up for a 7-acre (3-hectare) experimental farm. The United States Navy appointed him director of the project, and the

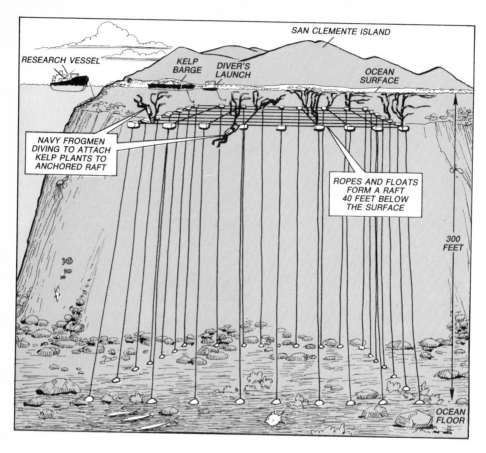

Ocean energy farm off San Clemente Island (Drawing by Russell Arasmith, ©, 1974, Los Angeles Times. Reprinted by permission)

first energy farm (somewhat smaller than originally planned) was installed in the Pacific Ocean in the spring of 1974.

This farm was located in the open ocean, 60 miles (96 kilometers) off the California coast, not far from San Clemente Island. The question was: Would kelp plants thrive if they were transplanted from shallow, nutrient-rich coastal water to deep, nutrient-poor oceanic

Close-up of a diver attaching kelp to the ropes (General Electric)

water? Furthermore, as everyone knows, plants need light to grow in, but the deeper the water, the darker it gets down there. On the deep, dark ocean bottom, transplanted young kelp wouldn't have much of a chance.

"So," said Wilcox, "we fooled the kelp into thinking the ocean was shallow. We built a large raft; crisscrossed it with polypropylene ropes, moored it 40 feet (12 meters) below the surface of the water, and anchored it with long lines to the ocean bottom 300 feet (90 meters) below."

Then Professor Wheeler North of the California Institute of Technology, who is the world's leading authority on kelp, entered the picture. North and a couple of assistants slipped into shiny black diving suits, masks, goggles, and fins. They shouldered their air tanks and scouted the water of a nearby coastal bed. There they selected about a hundred young kelp plants of the giant California variety, *Macrocystis pyrifera*. These were then transplanted onto the submerged raft by a team of Navy frogmen, turned tailors. The frogmen dived down to the raft and sewed the young slippery kelp onto the ropes with long darning needles specially designed by North in his environmental science office in Pasadena.

Macrocystis pyrifera is one of the most important seaweeds and grows almost as tall in the ocean as the giant redwoods grow on land. Like the redwoods, it grows in forests, but these are underwater forests. Here, however, the similarity ends. Redwoods absorb nutrients and water through their roots. Kelp have neither roots nor leaves. Instead, they have *holdfasts*, tiny tendrils that, under natural conditions, wrap around sunken rocks and in this way anchor the plant. As for nourishment, every part of the kelp takes care of that by absorbing water and other minerals.

Once a young kelp is anchored, it begins growing upward toward the light. When it reaches the surface of the water, its shiny brown fronds, supported by little air sacs, begin streaming through the sunlit sea like slippery ribbons. It's in these streaming fronds that solar energy is converted to chemical energy, in the process called photosynthesis.

Although Professor North had carefully supervised the transplant-
ing of kelp plants onto the raft in the Pacific, they did not perform
as expected. Periodically, naval personnel checked the mooring
buoys and navigational markers. And at regular intervals North

Dr. Wheeler North diving to check the kelp
on the ropes (Floyd Clark, Caltech)

dived down to the raft to check the ropes, as well as the kelp, which he found just struggling along. The plants were not growing vigorously because, as a chemical analysis indicated, the nutritional level of that blue water was low.

At this point you might ask, "What are the nutrients that might be lacking in the water of that bright blue ocean?"

Well, the color of the ocean tells its own story. Where the water is blue, the ocean is a "biological desert" unable to support more than a scattering of sea life. Since it is virtually devoid of life-sustaining nutrients, very few plants and animals can live there. Consequently, very few get to die and disintegrate there. So the water remains clear and blue, without organic matter and without certain necessary minerals that act as nutrients.

Where the water is green (as it is around the edges of the continents), it is crowded with living and feeding and dying life: diatoms and plankton, crustaceans and clams, seaweed and fish. The reason? Chlorophyll—the green coloring matter in plants that is basic to the photosynthetic process. Accordingly, the rivers that empty into the sea enrich the water with all kinds of dissolved organic and inorganic matter. This makes it possible for the floating algae and the great kelp beds to flourish there naturally. And flourish they do, all around the world in the cool coastal waters of the temperate zones. The best beds grow where the bottom waters *upwell*, move up to the surface. This happens along the shores of California and Peru and the other eastern shores of the Pacific.

For large kelp beds to flourish as farms in the biological desert of the open ocean, fertilizing agents have to be provided. These include nitrogen, phosphorus, and certain micronutrients. To some extent, these agents are contained in the deeper water, having accumulated there through the centuries as the remains of scattered plants and animals drifted down. For ocean energy farms, this deep water (which is actually enriched) would constitute almost free fertilizer if it were upwelled (pumped up) from the depths.

The Navy's ocean energy farm that Wilcox and North tended so diligently actually lasted less than a year, not long enough for the kelp to be harvested and converted to methane gas. In January 1975 disaster struck. A ship of unknown registry plowed through the farm, despite obvious marking buoys, and destroyed it. But no one could call the project a total loss. It proved the concept that coastal kelp could be transplanted onto a false bottom in the open ocean and not only survive but reproduce. Additionally, it demonstrated that the plants would require fertilizing before they could grow well in the open ocean.

About a year later, the Navy withdrew from its role as manager of the project. Support, however continued from two sources. General Electric, retaining Wilcox as consultant, took over and carried forward with all the original active agencies except for the Navy, which was replaced by Global Marine Development Inc. The United States Department of Energy provided additional funding and retained North as consultant.

Undismayed by the turn of events, Wilcox continued to urge the expansion of the ocean farm program in three phases. Following the original phase, his scenario called for two 1,000-acre (400-hectare) farms, one in the Atlantic and one in the Pacific. Later, if the work showed promise and the sponsors continued their backing, he envisioned as the third stage two 100,000-acre (40,000-hectare) farms, each about 12 miles (19 kilometers) square, again one in the Atlantic and one in the Pacific. According to his calculations, there are about 50 million square miles (130 million square kilometers) of "arable surface" in the world's oceans, and each square mile (which is equivalent to an area twenty city blocks long by twenty wide) may well yield enough fuel to support 300 people at today's United States energy consumption levels. "These farms," he pointed out, "would be well moored and fertilized by nutrient-rich water from deep below. And then we'd have no struggling kelp on those polypropylene ropes. And let's not forget the valuable byproducts: animal feed and petroleum-like commodities." He also notes another bonus—fish and other marine animals would thrive and increase in the upwelled nutrient-rich water around those farms.

As if all this were not enough to win popular support, Wilcox calls

A kelp harvesting ship in California waters (Kelco Company)

attention to the fact that no new technology is necessary. Proof: natural kelp beds are already being harvested by private firms such as the Kelco Company, and methane gas from raw chopped kelp has already been produced in the laboratories of the Institute of Gas Technology.

With customary enthusiasm but with practical restraint, Wilcox now urged the preliminary establishment of a ¼-acre (⅝-hectare) test farm, which in due time was installed in San Pedro Channel about 8 miles (12½ kilometers) from Corona Del Mar, south of Los Angeles. From this farm, Professor North was determined to learn to what extent fertilizer (in the form of nutrient-rich water) could be pumped up from the depths. And he was equally determined to find out to what extent this pumped-up water would stimulate the growth of kelp.

The quarter-acre test farm, like the original experimental farm off San Clemente Island, started with a raft that was submerged, moored, and anchored, but this raft was designed differently. It is an umbrella-like structure with the transplanted kelp, on the polypropylene ropes, encircling a central spar buoy. This central spar gives stability to the test farm. It also serves as the upper termination point of a 1,500-foot (450-meter) pipe to upwell the deep water with the aid of a diesel-powered pump. Established in 1978, this little test farm continues to serve as a base of study for Professor North.

A quarter-acre test raft being lowered for installation as an ocean energy farm off Corona Del Mar, California (General Electric)

A quarter-acre test energy farm 60 feet (18 meters) below
the Pacific off the California coast (General Electric)

When the time comes to proceed with the second phase of the
scenario, the development of the 1,000-acre plantations, the emphasis will be on reducing cost and boosting yield. As far as feasible,
these plantations would probably be patterned after the quarter-
acre farm. But the spar buoy would be topped by a platform with

living quarters for the ocean farmers. There would be designated areas for processing plants, for holding spaces, and for buoyancy and navigation controls. There might also be a wave-actuated pump for upwelling the deep water. And for a project this size, there would be a position-keeping propeller (called a propulser) so the plantation can be kept from drifting all over the ocean. A helicopter platform and a harvesting ship would also be included in the system.

Since kelp grows so rapidly, the harvesting ship would probably be in continuous operation, moving slowly over the sunken farm like a large aquatic lawnmower. The upper fronds of the kelp would be cut, hauled aboard, partly dewatered, and transferred to the processing plant.

In nature, the full growth of a kelp frond takes about six months. On a fertilized ocean farm the fronds would probably reach full growth in less than four months. And the yield could run as high as

Conceptual design of the 1,000-acre ocean food and energy farm
(U.S. Naval Undersea Center, San Diego)

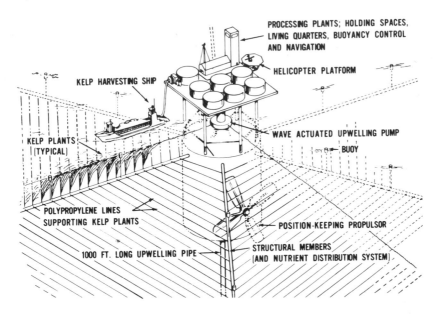

PROCESSING PLANTS; HOLDING SPACES, LIVING QUARTERS, BUOYANCY CONTROL AND NAVIGATION

KELP HARVESTING SHIP

HELICOPTER PLATFORM

KELP PLANTS (TYPICAL)

WAVE ACTUATED UPWELLING PUMP

BUOY

POLYPROPYLENE LINES SUPPORTING KELP PLANTS

POSITION-KEEPING PROPULSOR

1000 FT. LONG UPWELLING PIPE

STRUCTURAL MEMBERS (AND NUTRIENT DISTRIBUTION SYSTEM)

300 to 500 wet tons per acre. According to Professor North, "A kelp farm would not have to be replanted after each harvest because the plants would continue to grow after each cutting, just as hair does or as grass does after each mowing. Furthermore, since kelp naturally reproduces on the sunken raft, there would be a cancellation of loss if some plants happened to be torn away by storms, killed by disease, or damaged from excessive grazing by marine animals."

This then could be the ocean energy farm of the future. It would cover vast stretches of "submerged real estate." It would be marked by clanging, wave-powered buoys, flashing signal lights, and cruising harvest ships. It would be fertilized by nutrient-rich deep water upwelled from below. Such a farm could be nature's own energy converter in the bright blue ocean.

You will not be surprised to learn that Professor North spends most of his time setting up kelp nutrition experiments. When he's not at his chalkboard in Caltech, he heads for the Pacific at Corona Del Mar, where the ocean is his laboratory. In faded jeans, sneakers, and windbreaker, he points his motorboat to the west. Once in the open water, he quickly locates his upwelling pipes that, moored and anchored, are staked out at various depths—one at 500 feet, another at 1,000 feet, and still another at 1,500 feet (3 feet = 0.9 meter). Depending on the depth sample he needs, North ties up to the proper pipe, turns on a gasoline-powered pump, and fills his 5-gallon (19-liter) cans. These water samples he then lugs to his culture room in the Kerchoff Marine Laboratory.

That culture room, crowded with dripping hoses, sloshing aquariums, and humming pumps, has been the scene of countless experiments. Starting in the mid-1970s, North first compared the nutritive content of deep-ocean water with surface water by holding young *Macrocystis pyrifera* in 10-gallon (38-liter) aquariums equipped with deep narrow troughs, where the water could be circulated vigorously. The cultures showed that plants kept in surface water grew slowly while those immersed in water from the depths usually grew more rapidly.

North then decided to investigate possible nutritional deficiencies even in deep water. Was there, he wondered, one particular nutrient, or perhaps more than one, missing? To find out he ran a two-part series of experiments. In the first series, he used fresh seawater hauled from the depths and enriched with a mixture of his own making, which he called "a standard nutrient solution for *Macrocystis pyrifera*." This solution contained a number of trace minerals (familiar to those who read labels on vitamin-mineral bottles), including iron and iodine, manganese and zinc. In addition, he supplemented the mixture with molybdenum, cobalt, and arsenic. Into the aquariums then went the deep seawater, a measured amount of the standard nutrient solution, and some young kelp. A flip of the switch and the pumps hummed, the water circulated, and the experiment was on.

North carefully monitored not only the kelp in the aquariums that were enriched with the nutrient solution, but also the kelp in the aquariums filled with just plain seawater. He found that the kelp in the enriched water grew far better than the kelp in the plain seawater.

In part two of the experiment, young kelp plants were again grown in deep seawater that was enriched with the nutrient solution but with the omission of one trace element at a time. Now the results were quite variable. Sometimes withholding a certain nutrient boosted growth consistently; at other times, it didn't. Withholding certain other nutrients produced sickly plants, a condition that was irreversible and that no amount of additional nutrients could change.

Of course the chemical composition of seawater varies from time to time and place to place; consequently it may take years to find out which nutrient mix is best for kelp growth on any given farm. North knows this and finds it challenging. Challenging too is his work in the culture room with seeding experiments. He places the reproductive blades of the kelp—the sporophylls—in a tray of chilled water. This causes the blades to shed their spores onto a rope conveniently placed in the tray. The shed spores then attach themselves to the rope and begin to grow. Later, trained divers transplant the rope to underwater supports for further experimental treatment.

Why bother with these seeding experiments? The answer is that it costs much less to transplant a rope studded with young kelp than to send divers down to a moored and anchored raft in order to sew the young plants, one by one, onto the polypropylene lines.

It has long been common knowledge that kelp naturally flourishes in the cool coastal waters of the temperate zones. Since 1975, it has also been known that the large *Macrocystis pyrifera* species can be transplanted and cultivated in the cool waters of the open ocean. Now a question arises: Could this kelp, or a similar variety, or a hybrid created in the culture rooms be cultivated on energy farms in the warm, tropical oceans?

"Affirmative," say Dr. Wilcox, Professor North, and a number of marine biologists, "because even in the tropics cold, nutrient-rich water could be upwelled with pumps and pipes."

Not surprisingly, the development of kelp hybrids that would adapt to the tropics is on the research agenda of an increasing number of marine biologists. Also on the agenda is the development of kelp resistant to disease and pests, as well as kelp that will accept as-yet-untried nutrients. Additionally, some ocean engineers are concerning themselves with studies of ocean currents and natural upwelling areas, others with the engineering of rafts durable enough to withstand any kind of turbulence at 40, 60, or 100 feet (12, 18, or 30 meters) below the surface. Still others are investigating the possibility of coupling ocean energy farms with OTEC ships that also draw cold water from the depths.

When a good idea's time has come, research proliferates. Accordingly, ocean engineers and marine biologists in the United States, China, and Japan see a new era of plenty in the offing. They see an abundance of methane gas and ethanol processed at ocean energy farms and shipped by tanker to the mainland to fuel motorized vehicles, to heat and air-condition homes and schools, factories and office buildings. Writing in professional journals, addressing consumer groups, and lobbying in legislative bodies, these ocean scientists point out that, as the fossil fuel crunch worsens, governments all over the world should become more generous in their support of alternative forms of energy. They will find ocean energy farms particularly attractive for several reasons.

"Real estate" in the ocean is free, and it will continue to be free until governments find out how valuable it really is. Then they will probably assert their sovereignty over it, start to lease or sell it, and certainly tax it.

Irrigation problems, except for upwelling the deep water, are nonexistent.

Ocean energy farms won't pollute the planet thermally. The amount of energy that the growing kelp absorb from sunlight will, ultimately, be returned to the atmosphere after the methane gas, ethanol, and other products are burned or used up in other ways. The absorbed energy and the returned energy will cancel each other, and the earth's over-all thermal balance will remain exactly the same. The maintenance of the earth's thermal balance is a particularly important consideration today when so many scientists warn that the continued burning of fossil fuels will heat up the planet by perhaps two or three degrees F and bring about dire atmospheric changes.

As we arrive at the 1980s, the United States Department of Energy, General Electric, the American Gas Association, and Global Marine Development Inc., as well as university professors, environmentalists, and consumer groups who are sponsoring and supporting ocean energy farms, are convinced they are onto a good thing. They know that kelp is an excellent source of methane gas and ethanol. They know that the technology is available and the economics tempting. All of them, including the pioneers Howard Wilcox and Wheeler North, firmly believe that time will show such farms to be possible not only in principle but also in actuality.

8

Possible in Principle

The Wright brothers would never have gotten their flying machine into the air if, in 1903, they had had to start the way inventors do today, with submissions to the government of proposals and plans for design number one right on to the anticipated design for the full-scale model. And these plans would, in addition, have had to show technological soundness, economic feasibility, and environmental safety to a group of "peers," none of whom believed that flight was possible.

What the Wright brothers started with was only an idea for a flying machine—an idea that was possible in principle—and people laughed. People often laugh at what is possible in principle, because such ideas, being new and untried, sound altogether crazy. They laughed at Robert Fulton's "Folly" when it first steamed up the Hudson. They shouted, "Get a horse," when Henry Ford's first cars (the "tin Lizzies") got mired in the muddy roads. Even as late as July 19, 1969 (the day before Neil Armstrong leaped from Apollo 11 onto the lunar landscape), they scoffed at the thought of a man's walking on the moon.

As the twenty-first century approaches, the possible-in-principle spotlight swings again, this time to the coasts around the world. Here, flowing into the sea is the global runoff of creeks and streams, of mighty rivers and innumerable trickles after every rainfall. Here, at ocean's edge, two waters meet—one salty, the other fresh—and in meeting, naturally mix.

Now, whenever fresh and salt water mix, it's possible to extract energy from that mixing. Although more research is needed, calculations confirm the fact that the difference in salt concentration between river water and seawater is a tremendous potential source of energy.

Scientists are already "playing around" with simple energy-extraction devices that would convert this energy to mechanical or electrical power. They are developing salinity gradient energy converters (gradient meaning a gradation or change in salt concentration). These converters would operate on a basic physical principle called *osmosis.*

Osmosis is the process that permits a fluid (a liquid or a gas) to pass through a membrane that is selective and semipermeable. This means that the membrane, although permitting a liquid such as seawater to pass through, does not permit the passage of any salts that might be dissolved in the liquid.

Osmosis is a familiar but tricky phenomenon by means of which plants can absorb groundwater through their roots. You can see how osmosis works with reference to the mixing of river water and ocean water in a science laboratory or classroom where an osmotic tank is available. Such a tank is usually made of glass and divided into two sections by a vertically placed semipermeable membrane. If you fill the sections of such a tank one-quarter full, one with river water, the other with ocean water, you will soon see osmosis taking place. Pure water will begin to flow from the river section (which is only

The principle of osmosis

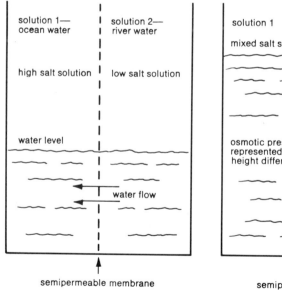

solution 1—
ocean water

solution 2—
river water

high salt solution

low salt solution

water level

water flow

semipermeable membrane

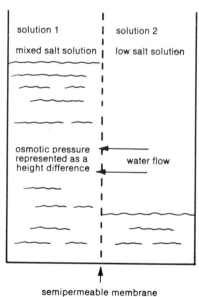

solution 1

mixed salt solution

solution 2

low salt solution

osmotic pressure
represented as a
height difference

water flow

semipermeable membrane

slightly salty) through the membrane into the saltier ocean section. This happens because, under normal conditions, the osmotic flow is always from the weaker salt solution to and into the more concentrated salt solution. This one-way flow will continue until both solutions are equal in salt concentration. By that time, of course, the ocean solution ("1" in the diagram) will have increased in volume and you will be able to see a noticeable difference in height between the two liquid surfaces (see second diagram on page 118). This creates a difference in pressure which scientists call osmotic pressure difference.

If you tried another experiment, filling each section of the tank three-quarters full with river water and ocean water respectively, you would see the water in the saltier section rise and rise until it spilled over the top of the tank. (See diagrams below.)

Osmotic pressure difference is a powerful force and will figure in the production of energy from salinity gradients, or, as some call it, salinity power.

In a number of marine laboratories, scientists are working to de-

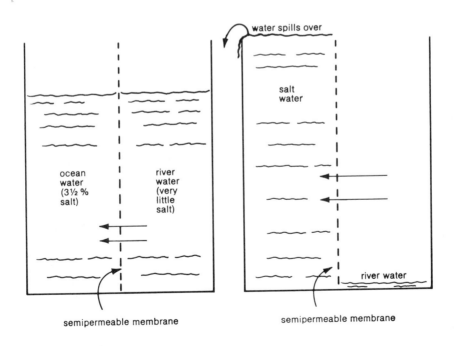

water spills over

salt water

ocean water (3½ % salt)

river water (very little salt)

semipermeable membrane

river water

semipermeable membrane

vise efficient and inexpensive osmotic converters that would be able to harness this salinity power so it would turn a waterwheel or spin a turbine and produce electricity.

At the Scripps Institution of Oceanography, professors Gerald Wick and John Isaacs believe that it would be possible, with osmotic converters, to extract as much energy at ocean's edge, where rivers flow into the sea, as is now extracted from all hydroelectric sources. And hydroelectric sources in the United States provide as much as 14 percent of all the electricity that's used.

Presently, however, this energy is totally lost. An example is the Columbia River. As much energy is lost at the mouth of this river as is now extracted by all of its hydroelectric dams. It is dissipated in the riotous mixing of the waters as the Columbia flows into the Pacific. Even more energy is dissipated at the mouth of the Mississippi. According to Richard Norman, professor of biological sciences at the University of Connecticut, a salination power plant at the mouth of that river, using only one-tenth of the flow, could deliver 1,000 megawatts of electricity, enough to serve 1,000,000 people there.

Because Norman sees this great energy potential graphically, he asks us to imagine every river in the world terminating on a cliff 750 feet (225 meters) above sea level. Every river then, would be a waterfall releasing as much energy as if it were plunging down a 750-foot bank. (Compare this with Hoover Dam, which is only 726 feet high.) Other scientists figure the height of these theoretical waterfalls between 760 and 798 feet (228 and 240 meters). Any of these figures is astonishing enough if you think of the potential energy in a column of water falling that distance. Since, however, there are no uniformly tall oceanside cliffs off which fresh-water rivers could plunge and turn turbines in power plants, salinity power scientists are looking for a comparable energy potential created by osmotic pressure differences.

Question: Where would you find such tremendous osmotic pressure differences?

Answer: Hundreds of feet below sea level, at ocean's edge where fresh-water rivers meet and mix with the ocean which, on the average, contains a salt concentration of $3\frac{1}{2}$ percent.

At the mouth of a given river, these scientists would go down 750, 760, 798 feet (take your choice), to the level at which the osmotic pressure difference is great enough to be harnessed to spin a turbine. There they would install a power plant, generate electricity, and transmit it on high-voltage lines to the nearest power grid.

Now harnessing an osmotic pressure difference of $\frac{1}{7}$ mile ($\frac{1}{5}$ kilometer) below the surface of the water is possible in principle, but it's easier said than done. At the mouth of a river, such a project would require two huge underwater dams, an artificial buffer lake between the dams, a long pipe bringing river water to the power plant (798 feet below sea level) and semipermeable membranes capping another pipe that goes through the second dam. Through those membranes, the used river water would be drawn out by osmosis to the salty sea.

In *Energy from the Ocean,* a 1978 United States government publication, professors Wick and Isaacs, front-runners in this area, describe this two-step process for an estuarine salinity gradient energy converter (estuarine technically refers to river mouth).

"In the first step, the river water is allowed to flow through hydroelectric turbines into a reservoir at a level which is some hun-

Estuarine salinity gradient energy converter
(Diagram by D. P. Shoemaker for Gerald L. Wick and John D. Isaacs, "Utilization of the Energy from Salinity Gradients")

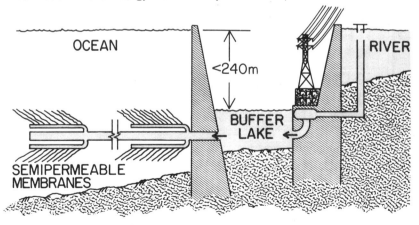

dreds of meters below sea level. The difference in heights is maintained by the osmotic pressure difference. Ideally it would be as much as 240 meters but it should be enough less than the osmotic equilibrium level (240 meters) to provide the driving force for the second step which follows. This second step is 'waste disposal.' It constitutes the discharge of the fresh water directly into the salt water through semipermeable membranes.... These membranes must be washed continuously by large volumes of ocean water in order to carry away the fresh water and prevent 'concentration polarization' due to dilution of salt water adjacent to the membrane surfaces."

Simply put, fresh water from the river enters a pipe in the first dam, turns a turbogenerator, and is exhausted into the lake. Then fresh water is "pumped out" to the salty sea through a pipe in the second dam. It's the membranes on this pipe that permit the pumping out, the drawing out, the sucking out, to take place. And the driving force behind it is the osmotic pressure difference of 240 meters, as Wick and Isaacs point out.

By 1979, a number of countries, led by Israel, Sweden, the United States, and Japan, had begun funding salinity power programs, and innovative scientists were flooding energy departments with proposals for electric generation based on osmotic pressure differences.

One such proposal receiving considerable attention is a converter for pressure-retarded osmosis (PRO). This technique is being investigated by Sidney Loeb in Israel's Ben-Gurion University of the Negev.

In Loeb's converter (see diagram on page 123), fresh water flows though a series of semipermeable membranes into a chamber containing seawater. But this is no ordinary seawater; it has been pressurized by a pump. Then the mix of pressurized seawater and osmotically added fresh water is directed to a turbine.

According to Loeb, the PRO converter is probably the most economical of the osmotic production processes for the generation of electricity. But he is quick to add, "Research and more research and

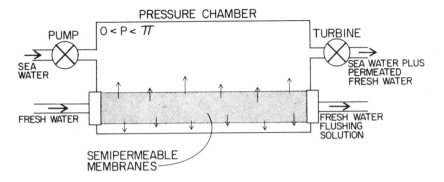

PRESSURE CHAMBER

PUMP $0 < P < \pi$ TURBINE

SEA WATER

FRESH WATER

SEA WATER PLUS PERMEATED FRESH WATER

FRESH WATER FLUSHING SOLUTION

SEMIPERMEABLE MEMBRANES

Schematic diagram of a "pressure-retarded osmosis" energy-conversion device (Gerald L. Wick and John D. Isaacs, "Utilization of the Energy from Salinity Gradients")

still more research will be necessary before this device, which looks so simple, is acceptable on a commercial basis in the marketplace."

Still another converter, a fresh-water/salt-water battery, intrigues the scientifically minded layman. This converter is electrochemical in nature and is in no way related to osmotic pressure differences. Being electrochemical, this battery is based on the movement of ions. (An ion is an atom that has gained or lost one or more electrons.)

The key to this process lies in the nature of salt—which is sodium chloride, expressed chemically as NaCl. When salt is dissolved in water, sodium and chloride ions are automatically formed. And these ions are restless. They move.

Two scientists, John Weinstein of the National Institute of Health and Frank Leitz of the United States Department of the Interior decided to capitalize on the movement of these ions to try to develop a fresh-water/salt-water battery. The idea was not totally new. It had been tried, reported on, and abandoned by scientists in the 1950s. Weinstein and Leitz followed up on those previous studies with an eye to possible extraction of energy on a commercial scale.

In actuality, the battery of the late 1970s is operated by the movement of the sodium and chloride ions through a stack of alternate

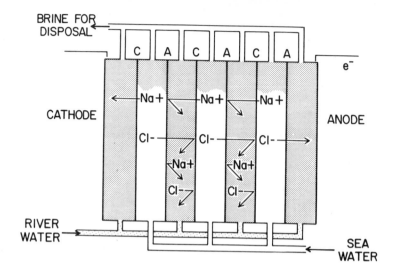

A salt-water/fresh-water battery can be made with alternate cells of ocean and fresh water separated by semipermeable membranes. The "A" refers to an anion permeable membrane and the "C" to a cation permeable membrane. Anion is a negatively charged ion, and cation is a positively charged ion. "Cathode" is the negative pole or electrode of an electrolytic cell; "anode" is the positive terminal of an electric source. (Reprinted with permission from *Popular Science*, © 1975 Times Mirror Magazines, Inc.)

cells of fresh water and salt water, ending in electrodes (terminals). Each of these cells is separated from its two neighboring cells by one of two kinds of membranes: one kind permits only the passage of the positive sodium ions (noted as Na+ on the diagram); the other kind permits only the passage of the negative chloride ions (noted as Cl− on the diagram). The positive ions zip one way, the negative ions zip the other way, and this orderly movement produces an electric current from an ingenious fresh-water/salt-water set-up.

Now back to salinity power for the generation of electricity, and a few more innovative proposals. Octave Levenspiel, professor of chemical engineering at Oregon's State University in Corvallis offers four of them. In personal communications as well as in professional papers, he describes and illustrates these proposals. He bases them all on the force of osmotic pressure differences. He maintains that all are possible in principle. (See diagrams on pages 126–127.)

One of his schemes calls for two underwater dams at the mouth of a river, another for one dam, a third only for the use of a large plastic sheet. "But my latest scheme is the best," he writes. "It needs no dams, no sheets, only a long pipe, a semipermeable membrane at the end of the pipe, and a turbogenerator."

With some modifications, this two-dam proposal of Levenspiel's resembles that of Isaacs and Wick and would produce electricity almost free of operational costs.

His one-dam scheme is just that. One large underwater dam (comparable in height to Hoover Dam) separates the ocean from the river and the mainland. A powerhouse, installed 700 feet (210 meters) below sea level is fed by a pipe from the river. "We lose the little lake," he says, "but the system will still work because the fresh water will still be osmotically sucked into the ocean. And look how much better and cheaper this system would be. Only one thin dam since we don't have to hold back two enormous walls of water."

In the third scheme, there are no dams. "We would," says Levenspiel, "just stretch a large plastic sheet across the mouth of the river down to a depth of 700 feet (210 meters). This would keep the fresh water and the salt water apart." (Of course there would be engineering problems; you can probably think of a few right now.) But the plastic-sheet plan is possible in principle. A 700-foot pipe would carry the river water down to the underwater power plant, and after the water was discharged from the power plant it would be directed through a relatively short exhaust pipe. From this pipe, which is capped by a semipermeable membrane, the fresh water would be osmotically sucked into the salty sea.

Levenspiel calls his fourth scheme the best of all plans. "It calls for no dams, no sheets, and no interference with boat traffic. It is just an intake pipe above the tidewater line. And remember, each particle of fresh water that flows through that long pipe down to the power plant carries the energy equivalent of a 760-foot dam."

An even more ingenious device offered by the resourceful Levenspiel is an osmotic pump. Of this pump (which has been described in

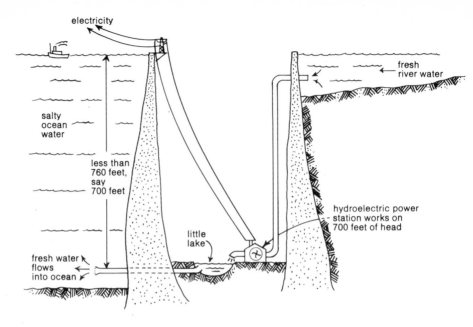

Extracting salinity power osmotically with a two-dam
underwater energy converter (Octave Levenspiel)

Extracting salinity power osmotically with a one-dam
underwater energy conventer (Octave Levenspiel)

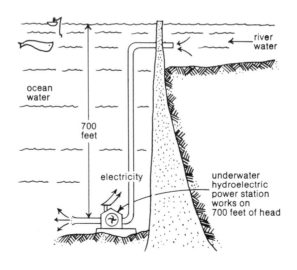

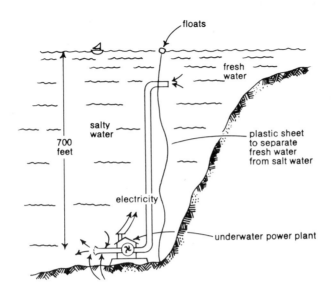

Extracting salinity power osmotically by way of a large plastic sheet (Octave Levenspiel)

Extracting salinity power osmotically by way of an intake pipe alone (Octave Levenspiel)

Scientific American, Science, Chemical Engineering, and a number of textbooks) he writes: "The osmotic pump is nothing more than a semipermeable membrane stretched across the bottom of a large pipe or tube. The membrane is permeable to water but not to dissolved salts. That's the pump. And it's completely scientific in principle." (See the diagram below.)

In practice, this astonishing pump, this long pipe capped with a semipermeable membrane, would be lowered deep into the ocean and anchored securely. Then, since this device depends on the fact that the flow of fresh water through the membrane is affected by the pressure difference on each side of the membrane, well, if the pipe were lowered less than 760 feet, the inside of the pipe would, at first, be free of water. If, however, it was pushed farther down, so that the membrane on the bottom was *below* 760 feet, then—surprise—the pressure of the ocean water would exceed the osmotic pressure. At such a depth, the process of osmosis would be reversed. Fresh water would be forced to flow through the semipermeable

The osmotic pump (Octave Levenspiel)

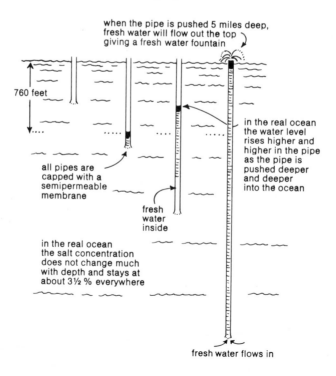

when the pipe is pushed 5 miles deep, fresh water will flow out the top giving a fresh water fountain

760 feet

in the real ocean the water level rises higher and higher in the pipe as the pipe is pushed deeper and deeper into the ocean

all pipes are capped with a semipermeable membrane

fresh water inside

in the real ocean the salt concentration does not change much with depth and stays at about 3½ % everywhere

fresh water flows in

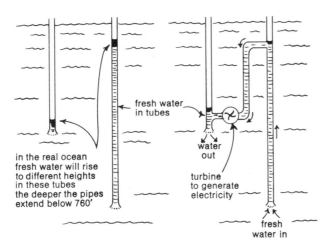

fresh water in tubes

in the real ocean
fresh water will rise
to different heights
in these tubes
the deeper the pipes
extend below 760'

water out

turbine to generate electricity

fresh water in

Using the osmotic pump for the generation of
electricity (Octave Levenspiel)

membrane in the opposite direction to normal osmotic flow. Consequently, fresh water would flow from the salty sea and begin to rise
in the pipe.

Now, if this is the way the osmotic pump works—if it can indeed
reverse the process of osmosis in the deep ocean—then one should
be able to get plenty of fresh water up to the surface of the sea—
fresh water to drink, to bathe in, to irrigate the deserts and make
them bloom.

"To get fresh water from the depths of the ocean," Levenspiel
says, "just get a longer pipe and push it down to a depth of 5 miles (8
kilometers). You'll not only get fresh water rising in your pipe, you'll
even see it bubbling out like a fountain. It's a well-known fact that
sea water is 3 percent denser than fresh water. So, after the level of
fresh water in the pipe rises to the level at which osmotic pressure is
reached (760 feet), it would be boosted above that level and bubble
out over the surface of the sea."

To use the osmotic pump for the generation of electricity (and
theoretically that too is possible) it would be necessary to lower two
pipes to different depths in the ocean. Fresh water will then rise to
different levels in these pipes. By letting the water flow "downhill"

from the longer pipe, through an underwater turbogenerator, to the shorter pipe, it should be possible to produce electricity.

Such imaginative proposals for converting salinity power to electric power are purely twenty-first-century stuff with great promise. But to succeed, these proposals require a basic and indispensable item: a semipermeable membrane that is rugged, long-lasting, economical, virtually maintenance-free, and not likely to get fouled up with salt and marine organisms. To date, the best suggestion is that engineers use membranes made of hollow fibers about the thickness of a human hair. These fibers take a lot of pressure and have enormous surface area. They are already being used in artificial kidneys, and a bundle of 13,000 (made by Dow Chemical Company) is only about 2 inches (5 centimeters) in diameter. For these fibers, fouling would be no great problem in the ocean because, say some scientists, the ends of the tubes could be made soft and flattened and flushed out periodically.

It's clear that years of research and development (at least ten or fifteen) will be required before ideal semipermeable membranes are available at low cost, before biofouling can be handled with dispatch, before the technology is investigated and advanced. And let's

Paolo Soleri's conception of an ocean city (Photo by Ivan Pintar. Reprinted by permission of the MIT Press from Paolo Soleri/*Arcology:*

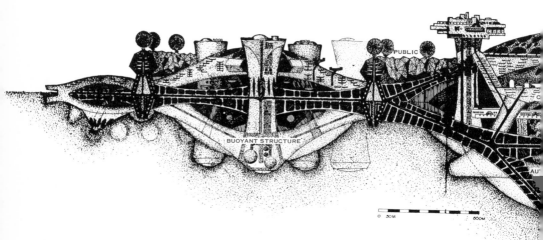

not forget the environment. Who can tell at this time what environmental impact might result from converters cutting into the mixing waters at the mouths of rivers?

But the boldest and most visionary of all possible-in-principle proposals, one that's based neither on salinity gradients nor on electrochemical processes, is Paolo Soleri's Ocean City. Soleri, architectural experimenter, urban planner, proponent of human ecologies, and president of the Cosanti Foundation, Scottsdale, Arizona, would take advantage of the energy in the flowing currents and the thermal disequilibrium of the seas. His Ocean City would have no fixed address but would drift with the currents or be slowly propelled by them on journeys according to the seasons, the sea harvests, and the fishing opportunities. And all the time the city would be served by OTEC power plants that would produce low-cost electricity and fresh water.

Science fiction or scientific fact? This won't be known for some time, but it is well to remember that schemes which are possible in principle today may be everyday happenings in the future. Remem-

The City in the Image of Man, copyright © 1974 by the Massachusetts Institute of Technology)

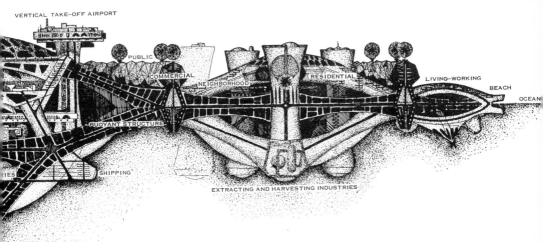

VERTICAL TAKE-OFF AIRPORT

PUBLIC

COMMERCIAL

NEIGHBORHOOD

RESIDENTIAL

LIVING-WORKING

BEACH

OCEAN

BUOYANT STRUCTURE

IES

SHIPPING

EXTRACTING AND HARVESTING INDUSTRIES

ber, too, that at the turn of the twentieth century, no one worried about principles or about energy development. Such words as energy were not in the vocabulary of the general public. The Western world was well into the scientific age, but the energy needed came from wood and water, wind, and some black mountain rocks called coal. These energy sources were familiar, seemingly inexhaustible, and cheap, and the technology (another word not yet in common use) was simple.

Then oil and natural gas were pushed forward as alternative sources of energy. John Doe was not impressed. "Strange, foul-smelling stuff," he said. "O.K., O.K., so it burns faster and hotter, but who needs it? And how would you get it out of the ground?"

On a more sophisticated level, investors asked, "What about technology? And how can we raise the money for those monstrous rigs?"

In short order, the technology was developed so successfully that oil and natural gas replaced wood and coal in stoves and furnaces. Then windmills and waterwheels became obsolete because the fossil fuels were dirt cheap—with gasoline at two gallons for a quarter.

And so everybody went riding in cars and trucks, buses, vans, and airplanes, on tractors and motorcycles.

But suppose these alternative fuels had *not* been discovered and developed? Then the energy in wood and coal and wind and water would have been exploited with ever more sophisticated technology, would have been converted on a much larger scale to electric power. And civilization would probably have come to be much the same. We might even have landed a man on the moon. Apollo 11 was fueled, not by gasoline, but by hydrogen. Future historians might then have remembered the twentieth century for its largely unpolluted and uncontaminated environment.

Now imagine a change of scenario. Imagine that oil and natural gas were discovered for the first time only toward the *end* of the twentieth century, after solar energy on land and sea had been developed. Investors would probably have pushed for the exploitation of those alternative forms with their high energy density, but the public would have been aghast. Consumers would have protested. Lobbyists would have campaigned against those dirty fossil fuels.

But, in the last decades of the twentieth century, the public was not being asked to change *to* fossil fuels; it was asked to consider changing to alternative forms of safe and nonpolluting forms of energy, with solar and ocean energy heading the list.

The idea of changing from one product to another, even from one style to another, or one job to another, is challenging to some people but threatening to many others. The idea of changing to alternative forms of energy raised a multitude of scientific and financial questions. These stirred up disturbing but at the same time immobilizing emotions. "Because," says Walter Schmitt, specialist in oceanography at the Institute of Marine Resources in La Jolla, "in the matter of energy, the Western world deals in dreams. It believes so strongly in the 'technological fix' that problems of scarcity, price, and pollution are not as distressing as they might be because 'technology will fix it.' "

On the eve of the ninth decade of the century, ocean scientists are asking us to stop dealing in dreams and relying on the technological fix. They are calling attention to the ocean, which is charged with energy that flows from outer space—energy that is available, abundant, safe, nonpolluting, and renewable. These same scientists point to the many alternative forms in the ocean (of which only six have been considered in this book: the tides and waves, the ocean currents and thermal gradients, marine farms and salinity gradients). "The total energy potential of the ocean is enormous," they say, "but this energy is diffuse and can be utilized economically only in those geographic areas where it is concentrated. But where it is concentrated, there's enough to meet local demands, either alone or in combination with other energy sources."

The story of energy is an endless story, with countless alternative forms available for our use provided we can figure out ways to develop them efficiently and economically. But no matter how we view the need for energy, the sources of energy, and the quality and pricing of energy, we can only agree with the unknown pundit who said, "There are no simple solutions—only intelligent choices."

Bibliography

Books

Carr, Donald E. *Energy and the Earth Machine.* New York: Norton, 1976.

Clark, Wilson. *Energy for Survival: The Alternative to Extinction.* Garden City, N.Y.: Doubleday (Anchor Books), 1975.

DiCerto, Joseph J. *The Electric Wishing Well.* New York: Macmillan, 1976.

Gray, T. K., and Gashus, O. J., eds. *Tidal Power.* New York: Plenum Press, 1972.

Halacy, D. S., Jr. *Earth, Water, Wind, and Sun: The Energy Alternatives.* New York: Harper & Row, 1977.

Hammond, Allen L.; Metz, William D.; and Maugh, Thomas H. *Energy and the Future.* Washington, D.C.: American Association for the Advancement of Science, 1974.

Krauss, Robert W., ed. *The Marine Plant Biomass of the Pacific Northwest Coast: A Potential Economic Resource.* Corvallis: Oregon State University Press, 1978.

Meadows, Donnella H.; Meadows, Dennis L.; Randers, Jorgen; and Behrens, W. W. *The Limits to Growth: A Report for the Club of Rome's Project on the Predicament of Mankind.* 2nd ed. New York: Universe Books, 1974.

Metzger, Norman. *Energy: The Continuing Crisis.* New York: Crowell, 1976.

Ruedisili, Lon C., and Firebaugh, Morris W., eds. *Perspectives on Energy and the Environment: Issues, Ideals, and Dilemmas.* New York: Oxford University Press, 1975.

Wilcox, Howard A. *Hothouse Earth.* New York: Praeger (A Frank E. Taylor Book), 1975.

United States Government Publications

Energy from the Ocean. Report prepared for the Subcommittee on Advanced Energy Technologies and Energy Conservation Research, Development and Demonstration of the Committee on Science and Technology, U.S. House of Representatives, April 1978.

International Passamaquoddy Tidal Power Project and Upper Saint John River Hydoelectric Development. U.S. Department of the Interior, August 1964.

Ocean Systems: Program Summary. U.S. Department of Energy, February 1979.

Renewable Ocean Energy Sources: Part 1, OTEC. Office of Technology Assessment, May 1978.

Solar Energy Update: Abstracts 1696-2171. U.S. Department of Energy, May 1978.

Stewart, Harris B., Jr., ed. *Proceedings of the MacArthur Workshop on the Feasibility of Extracting Useable Energy from the Florida Current.* National Oceanic and Atmospheric Administration, Atlantic Oceanographic and Meteorological Laboratories, Miami, Florida, February 27-March 1, 1974.

Tidal Power Study for Cobscook Bay, Maine. U.S. Army Corps of Engineers, New England Division, 1978.

Articles

Anderson, J. Hilbert. "The Sea Plant—A Source of Power, Water, and Food without Pollution." Sea Solar Power, Inc., pp. 1–12.

Bamberger, C. E., and Braunstein, J. "Hydrogen: A Versatile Element." *American Scientist.* Vol. 63 (July-August 1975), pp. 438–447.

Bernstein, Lev B. "Energy and the Civil Engineer: Russian Tidal Power Station is Precast Offsite, Floated into Place." *Civil Engineering,* April 1974, pp. 46–49.

Boyle, Patrick. "Could Gulf Stream Fans Cool Florida?" *Los Angeles Times*, January 14, 1979.

"Britannia Measures the Waves from Afar." *New Scientist* (London), June 29, 1978.

Davies, Owen. "Seven Sources to Help Power Your Future." *Science Digest* (special issue on energy), October 1977.

Fisher, Arthur. "Energy from the Sea, Part II." *Popular Science*, June 1975, pp. 78–83.

_____. "Energy from the Sea, Part III, Marine Farms and Salt Water Batteries." *Popular Science*, July 1975, pp. 62–65, 116.

Flowers, Ab, and Bryce, Armond. "Energy from Marine Biomass." *Sea Technology*, October 1977, pp. 1–4.

Fuller, R. D. "Ocean Thermal Energy Conversion." *Ocean Management 4*, Lockheed Missiles and Space Co., 1978, pp. 243–258.

Glendenning, I. "Ocean Wave Power." *Applied Energy*, Vol. 3, No. 3 (July 7, 1978), pp. 206–217.

Griffin, Owen W. "Power from the Oceans' Thermal Gradients." *Sea Technology*, August 1977, pp. 11–15.

Holland, M. B. "Power from the Tides." *The Chartered Mechanical Engineers Journal* (London), July 1978, pp. 33–39.

_____. "Power from the Waves." *The Chartered Mechanical Engineers Journal* (London), September 1978, pp. 41–47.

Kohl, Jerome, ed. "Energy from the Oceans, Fact or Fantasy?" Conference Proceedings, North Carolina University, January 27–28, 1976, pp. 5–8.

Land, Thomas. "Europe to Harness the Power of the Sea." *Sea Frontiers* (International Oceanographic Foundation), Vol. 26, No. 6 (November–December 1976), pp. 346–349.

Levenspiel, Octave, and de Nevers, Noel. "The Osmotic Pump." *Science*, Vol. 183 (January 18, 1974), pp. 157–161.

Loeb, Sidney. "Osmotic Powerplants." *Science*, Vol. 189, No. 4203, (August 22, 1975), pp. 654–655.

"Methane from Seaweed." *Grid* (Gas Research Institute) Vol. 2, No. 1 (January 1979), pp. 1–3.

Mulcahy, Michael. "OTEC—from $85,000 to $35 Million in Six Years." *Sea Technology*, August 1977, pp. 16–18.

Norman, Richard S. "Water Salination: A Source of Energy?" *Science*, October 1974, pp. 350–352.

Oceanus (Woods Hole Oceanographic Institution), Vol. XVII, No. 5, Summer 1974 (entire issue).

Othmer, F., and Roels, Oswald A. "Power, Fresh Water and Food from Cold Deep Waters." *Science*, Vol. 182, No. 4108 (October 12, 1973), pp. 121–124.

"Reassessment of Fundy Tidal Power." *Bay of Fundy Tidal Power Review*, 1977, pp. 1–57.

"Research and Development Program to Assess the Technical and Economic Feasibility of Methane Production from Giant Brown Kelp." Sponsored by the American Gas Association, pp. 1–22.

Salter, S. H. "Wave Power." *Nature* (London), Vol. 249, No. 5459 (June 21, 1974), pp. 720–724.

Shaw, T. L. "An Appraisal of the Tides as an Alternate Energy Source." *Physics Education*, Vol. 13. London: The Institute of Physics, 1978, pp. 312–318.

Sheets, Herman E. "Power Generation from Ocean Currents." *Naval Engineers Journal*, April 1975, pp. 47–56.

Strong, C.L. "The Amateur Scientist." *Scientific American*, Vol. 225, No. 6 (December 1971), p. 100.

Subrahmanyam, K. S. "Tidal Power in India." *Water Power and Dam Construction*, June 1978, pp. 42–44.

Von Arx, William S. "Energy: Natural Limits and Abundance." Reprinted from American Geophysical Union, Vol. 55, No. 9 (September 1974), pp. 828–832.

Whitmore, William F. "OTEC: Electricity from the Ocean." *Technology Review*, October 1978, pp. 2–7.

Wick, Gerald L. "Power from Salinity Gradients." *Energy*, Vol. 3, No. 1 (February 1978), pp. 95–100.

Wick, Gerald L., and Isaacs, John D. "Salt Domes: Is There More Energy Available from Their Salt Than from Their Oil?" *Science*, Vol. 199, No. 4336, (March 31, 1978), pp. 1436–1437.

Wick, Gerald L., and Isaacs, John D. *Utilization of the Energy from Salinity Gradients*. Presented at the Wave and Salinity Gradient Energy Conversion Workshop, University of Delaware, May 24–26, 1976, sponsored by the U.S. Energy Research and Development Association. La Jolla, California: Institute of Marine Resources, Scripps Institution of Oceanography, University of California, 1978.

Wick, Gerald L., and Schmitt, Walter R. "Prospects for Renewable Energy from the Sea." *MTS Journal* (Marine Technology Society, Inc.) Vol. 11, Nos. 5 and 6, 1977, pp. 16–20.

Wilcox, Howard A. "Farming the Sea." *Solar Age*, August 1976, pp. 6–10.

———. "The Ocean Food and Energy Farm Project." Supplement to *Calypso Log* (Cousteau Society, Inc.), Vol. 3, No. 2, 1976, pp. 1–6.

Zenner, Clarence. "Solar Sea Power." *Physics Today*, January 1973, pp. 48–53.

Papers

Behlke, Charles E., and Carlson, Robert F. "An Investigation of Small Tidal Power Plant Possibilities on Cook Inlet, Alaska." University of Alaska, April 1976.

Bryce, Armond J. "A Review of the Energy from Marine Biomass Program." General Electric Company, 1978.

Cohen, Robert. "An Overview of the U.S. OTEC Development Program." U.S. Department of Energy, July 15, 1978.

Leishman, J. M., and Scobie, G. "The Development of Wave Power: A Techno-Economic Study." National Engineering Laboratory, East Kilbride, Glasgow, Scotland, 1976.

Lissaman, Peter B. S. "Ocean Turbines: An Effective System for Current Energy Extraction." Symposium on Energy and the Oceans, Institute on Man and the Oceans, Inc., Miami, Florida, September 1977.

North, Wheeler J. "Progress in Studies of Oceanic Production of Biomass." Second Annual Symposium on Fuels from Biomass, Rensselaer Polytechnic Institute, Troy, N.Y., June 22, 1978.

North, Wheeler J., and Wheeler, Patricia A. "Nutritional Requirements of Macrocystis." Ninth International Seaweed Symposium, Santa Barbara, California, August 21, 1977.

Sheets, Herman E. "OTEC—A Survey of the Art." Presented for the Society of Naval Architects and Marine Engineers, Star Symposium, San Francisco, California, May 1977.

"Wave Energy." Department of Energy, paper number 42. London: Her Majesty's Stationery Office, 1979.

Wayne, W. W., Jr. "The Current Status of Tidal Power: Can It Really Help?" Symposium on Energy and the Oceans, Miami, Florida, October 31, 1977.

Wilcox, Howard A. "Artificial Oceanic Upwelling." Naval Undersea Center, San Diego, California, July 1975.

Index